# 2nd and 3rd Grades Multiplication Tables 1-12 with Exercises

Marina Vigodsky

Copyright © 2018 Marina Vigodsky
All rights reserved.
ISBN: **1727535642**
ISBN-13: **978-1727535648**

# Introduction

The "2nd and 3rd Grades Multiplication Tables 1-12 with Exercises" workbook helps 2nd and 3rd graders learn multiplication.

Multiplication tables are followed by multiplication drills which provide practice for 2nd and 3rd graders studying multiplication.

Visual memory facilitators used in two multiplication teaching buttons in this workbook make multiplication learning visually available. The first multiplication button is a three-sector button with factors and a product. The second button for teaching multiplication is a recall and rehearsal button with a blank space inside for writing products. Being connected with the first button it contributes to improved recall. Two buttons make a teaching cycle "introduction - rehearsal - recall" and work as a unit in teaching multiplication. This teaching unit provides a framework for improved introduction, rehearsal and recall of multiplication factors and products. From cognitive standpoint the contextual connection between two teaching buttons contributes to efficacy of multiplication drills and memorization of multiplication tables.

# One times table

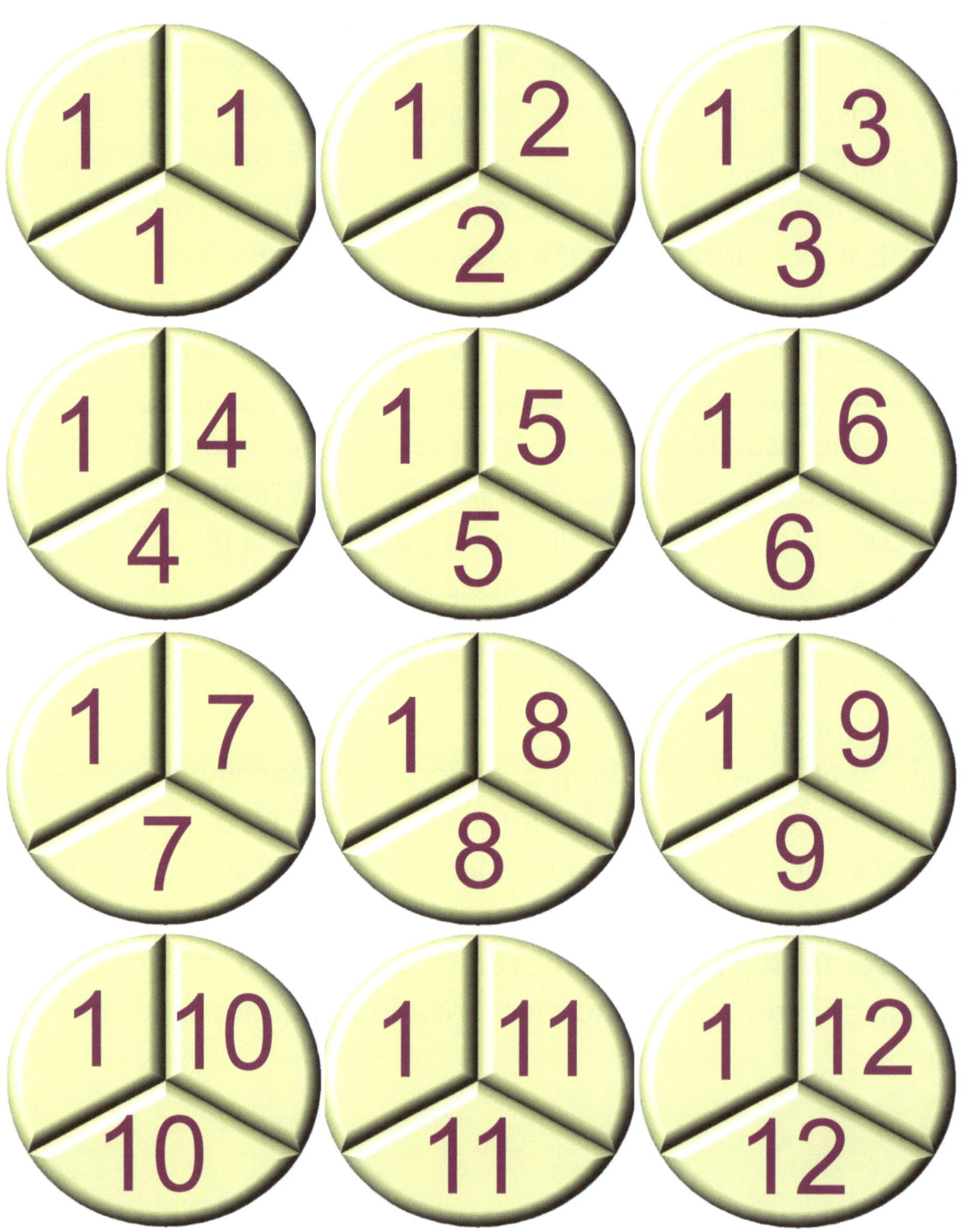

# One times drills

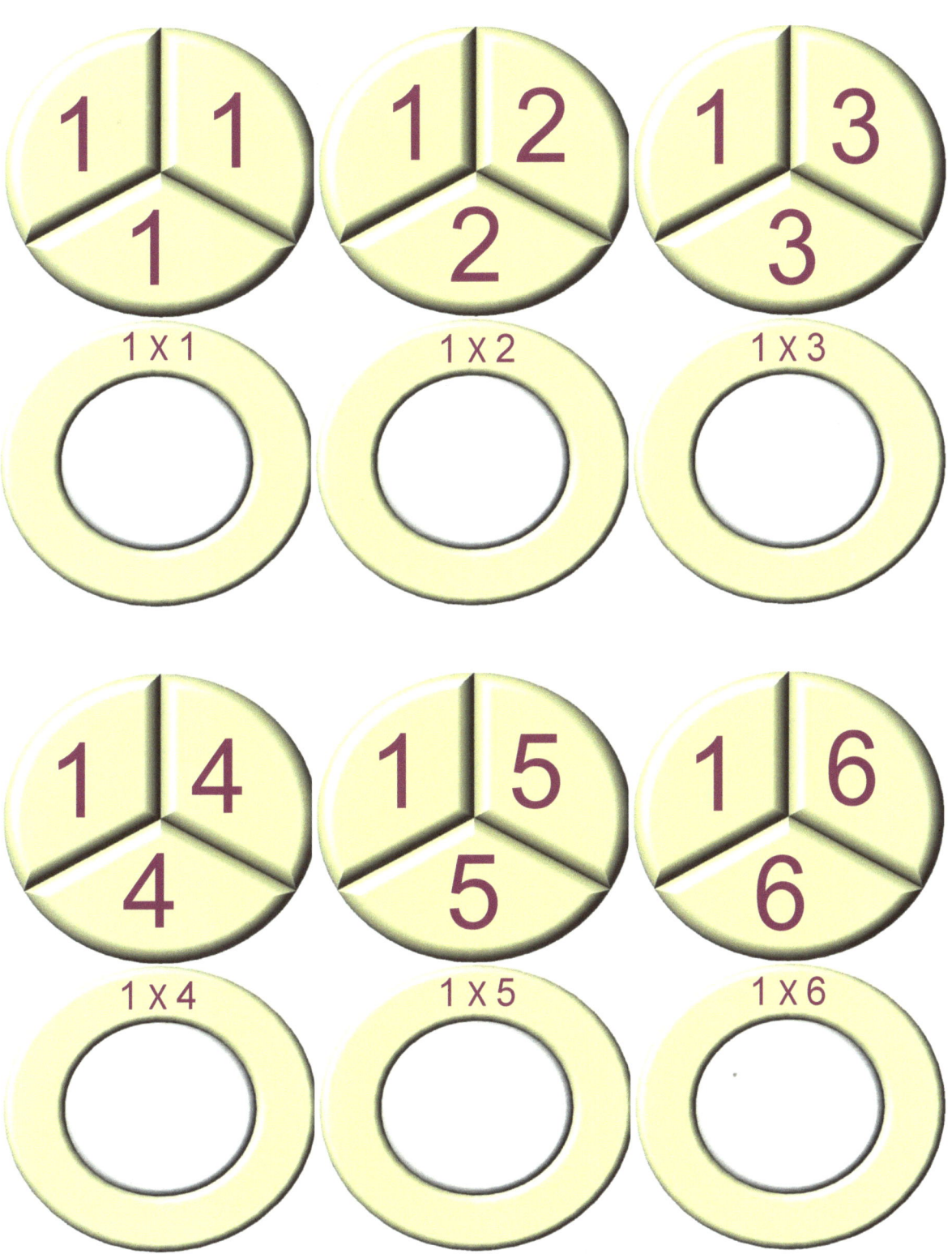

# One times drills

# Two times table

# Two times drills

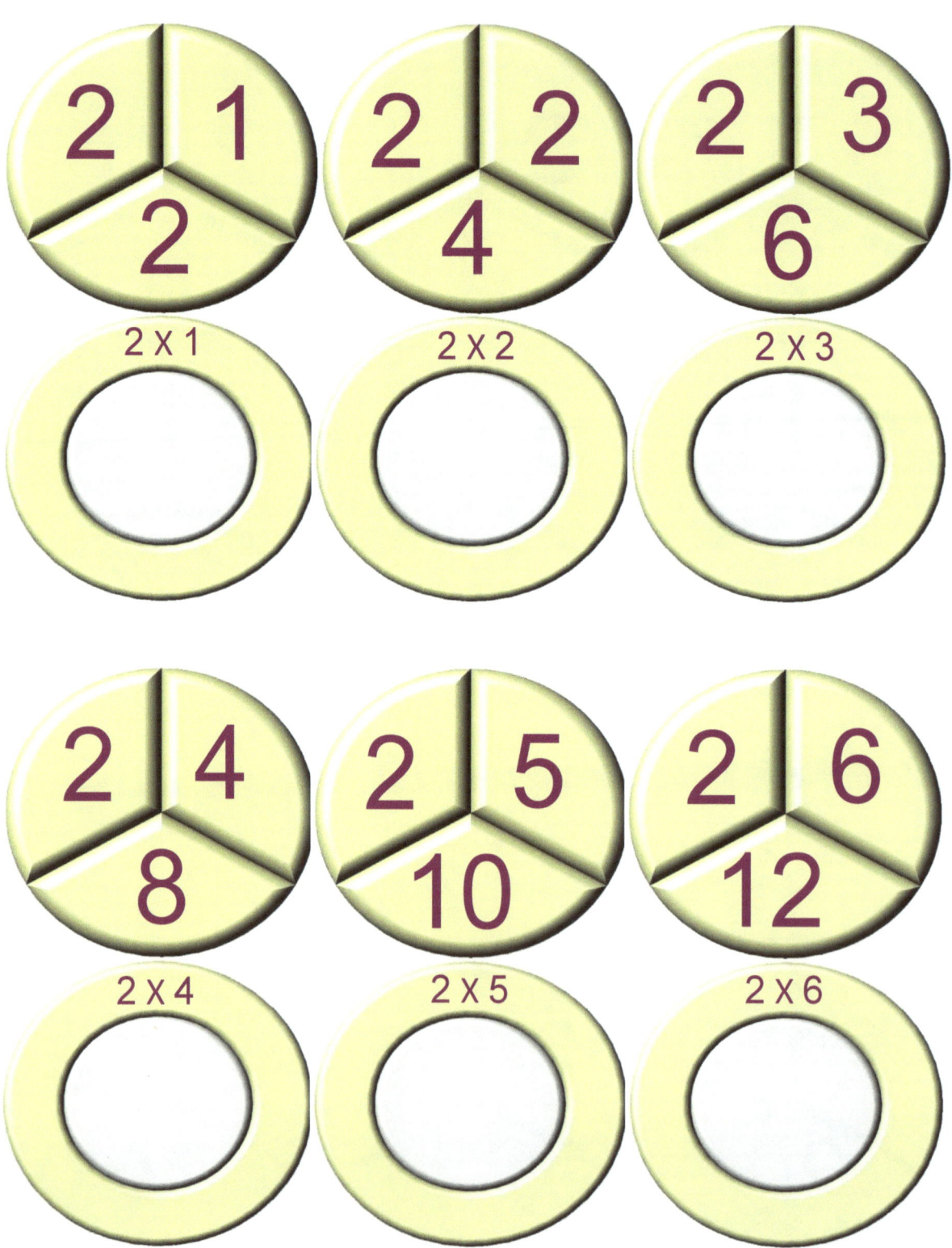

# Two times drills

# Three times table

# Three times drills

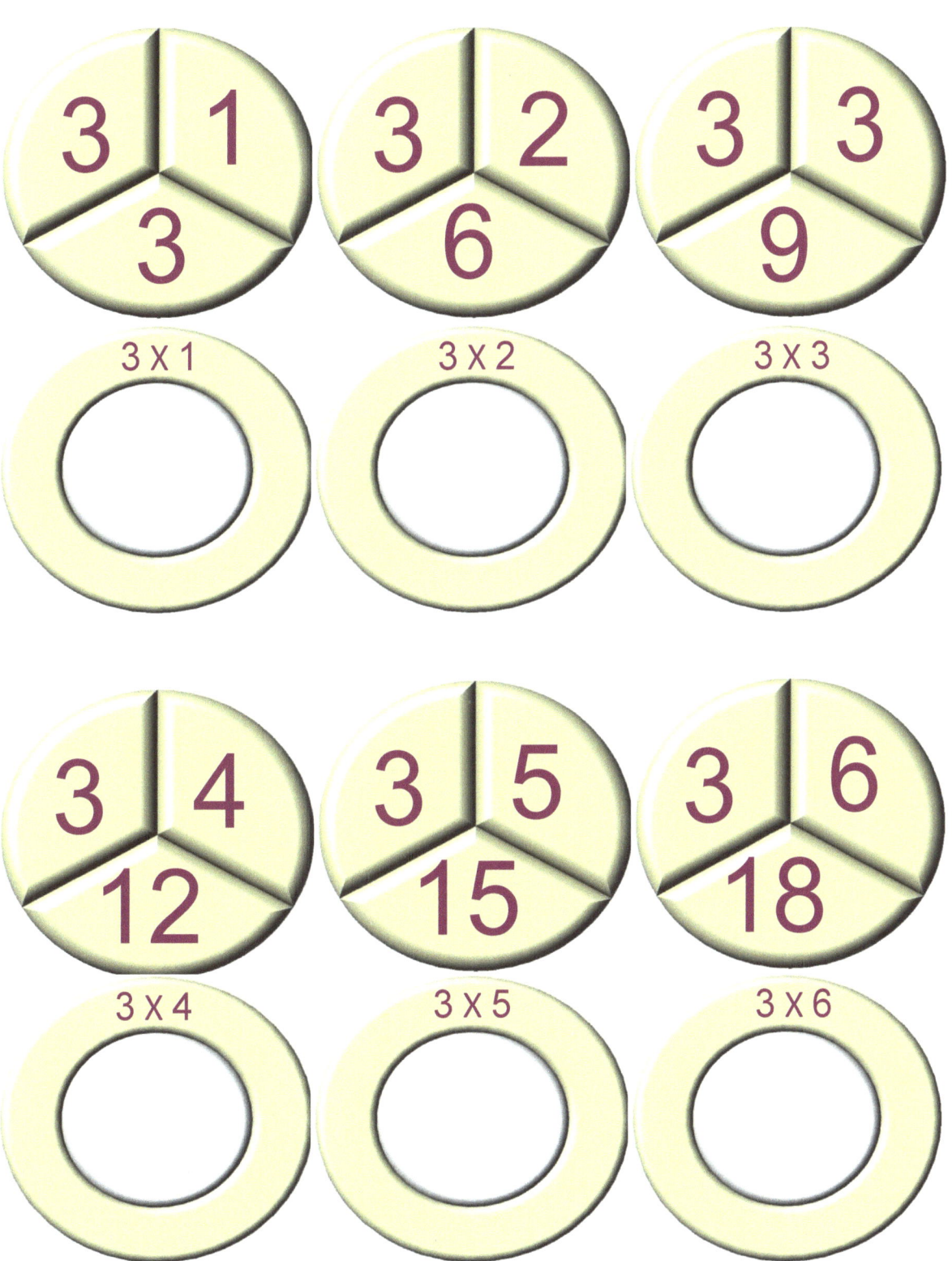

# Three times drills

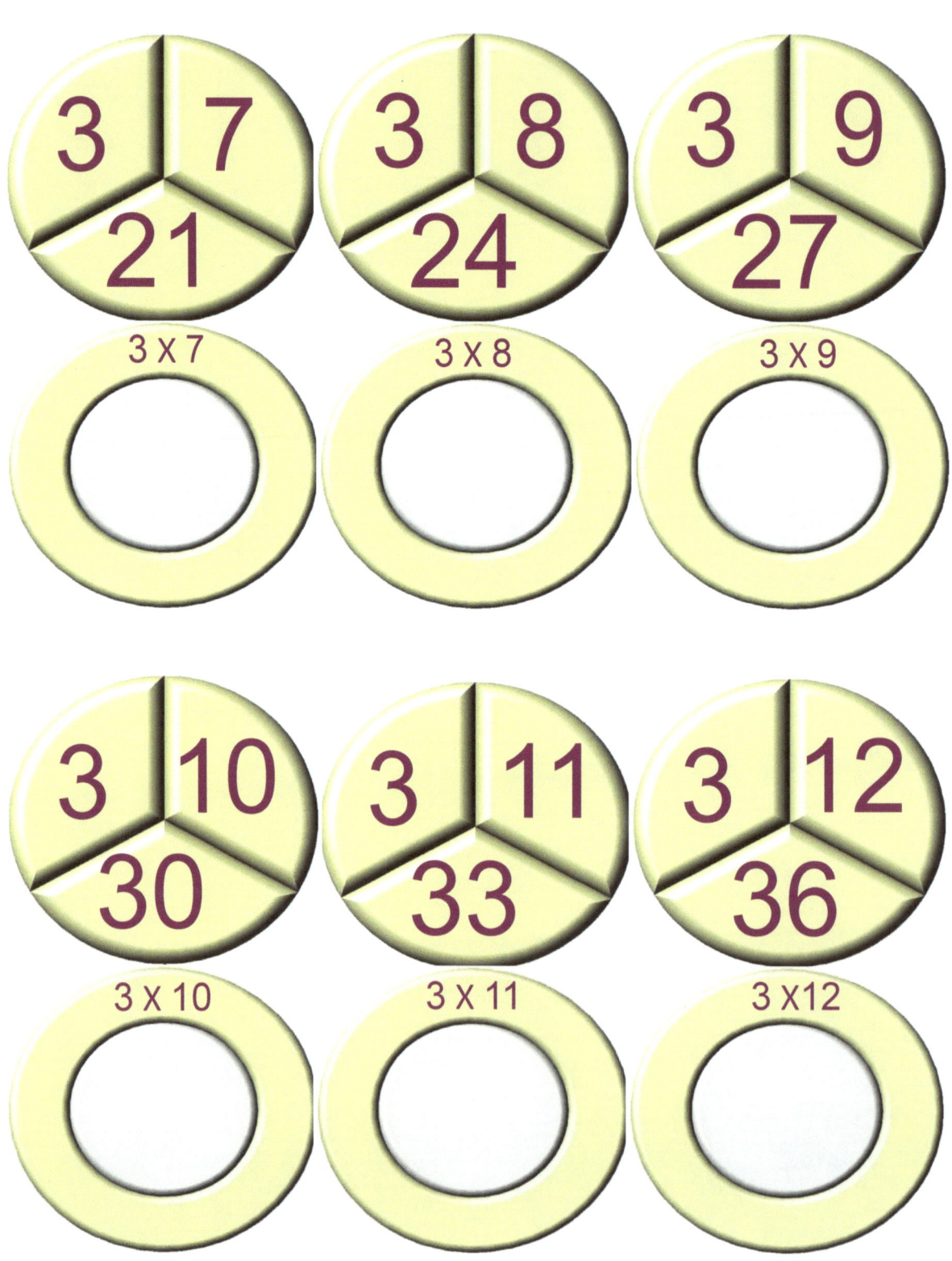

# Four times table

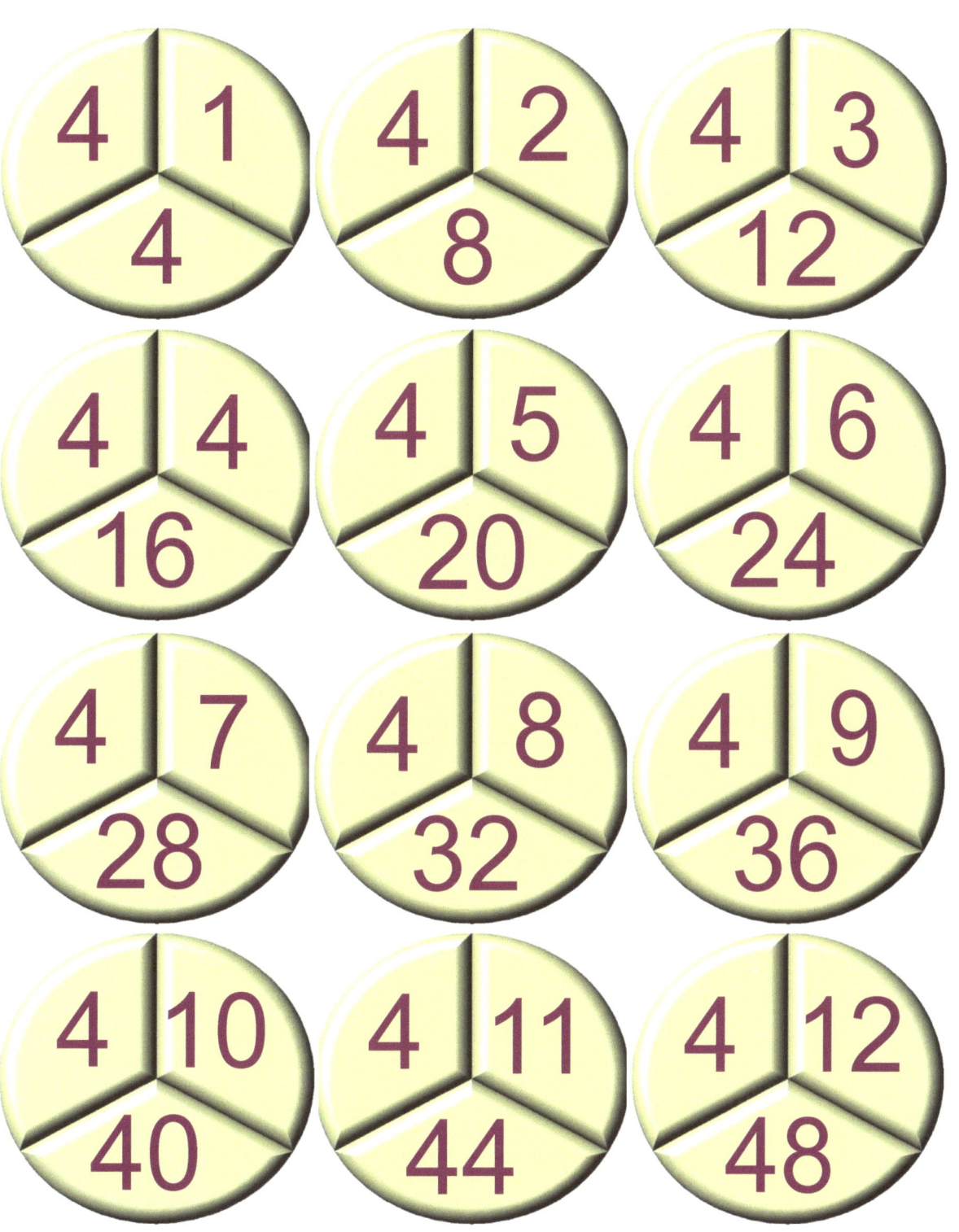

# Four times drills

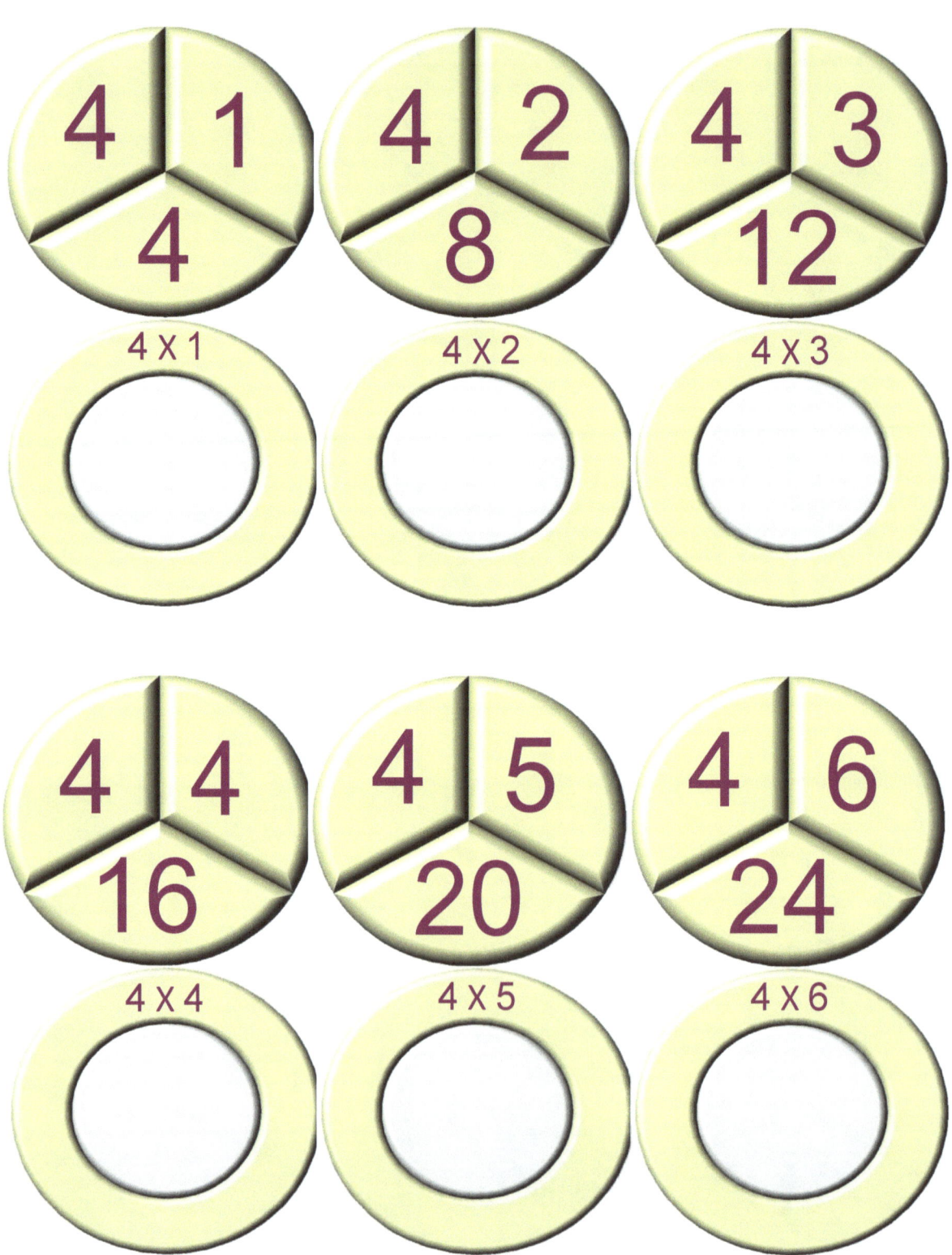

# Four times drills

# Five times table

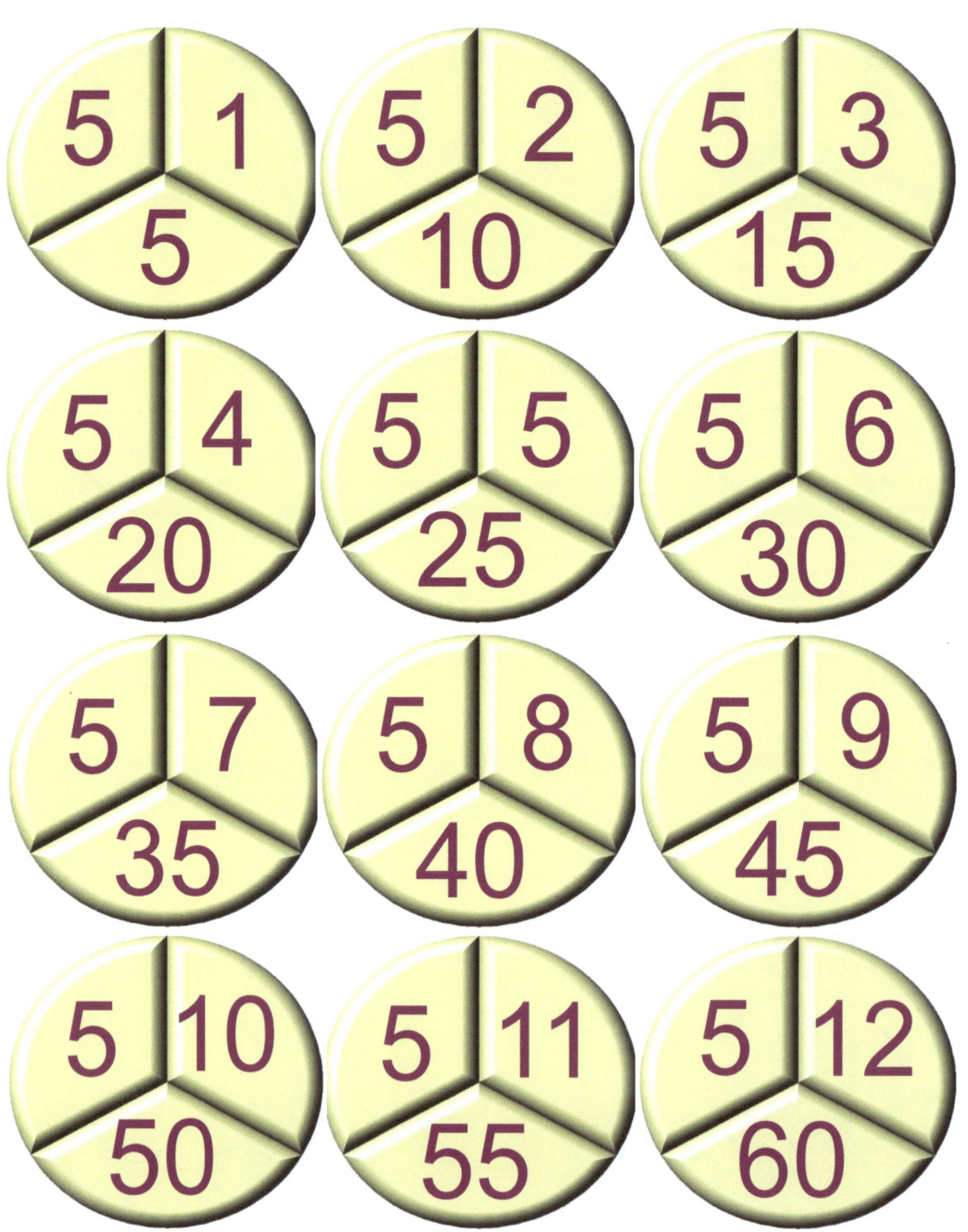

# Five times drills

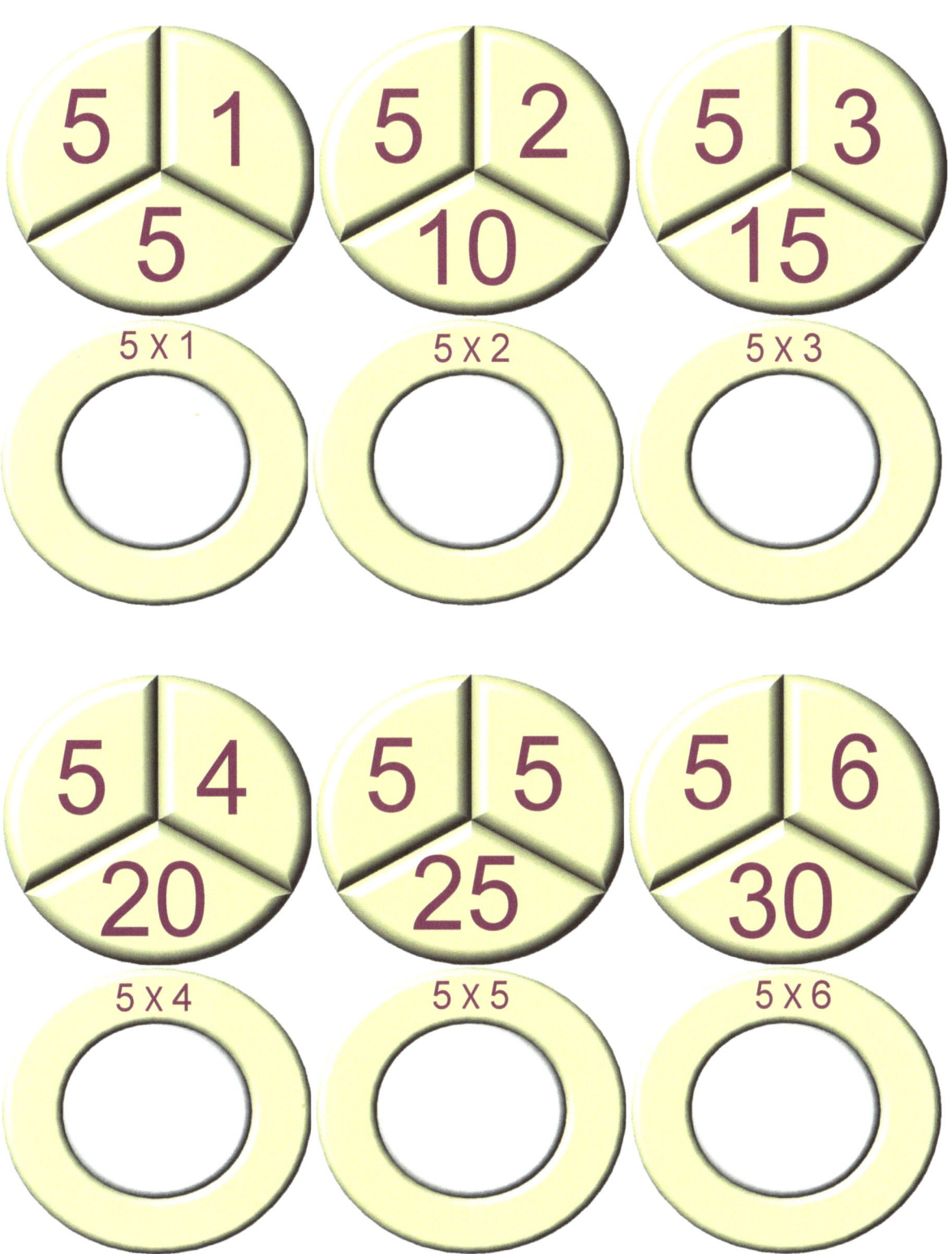

# Five times drills

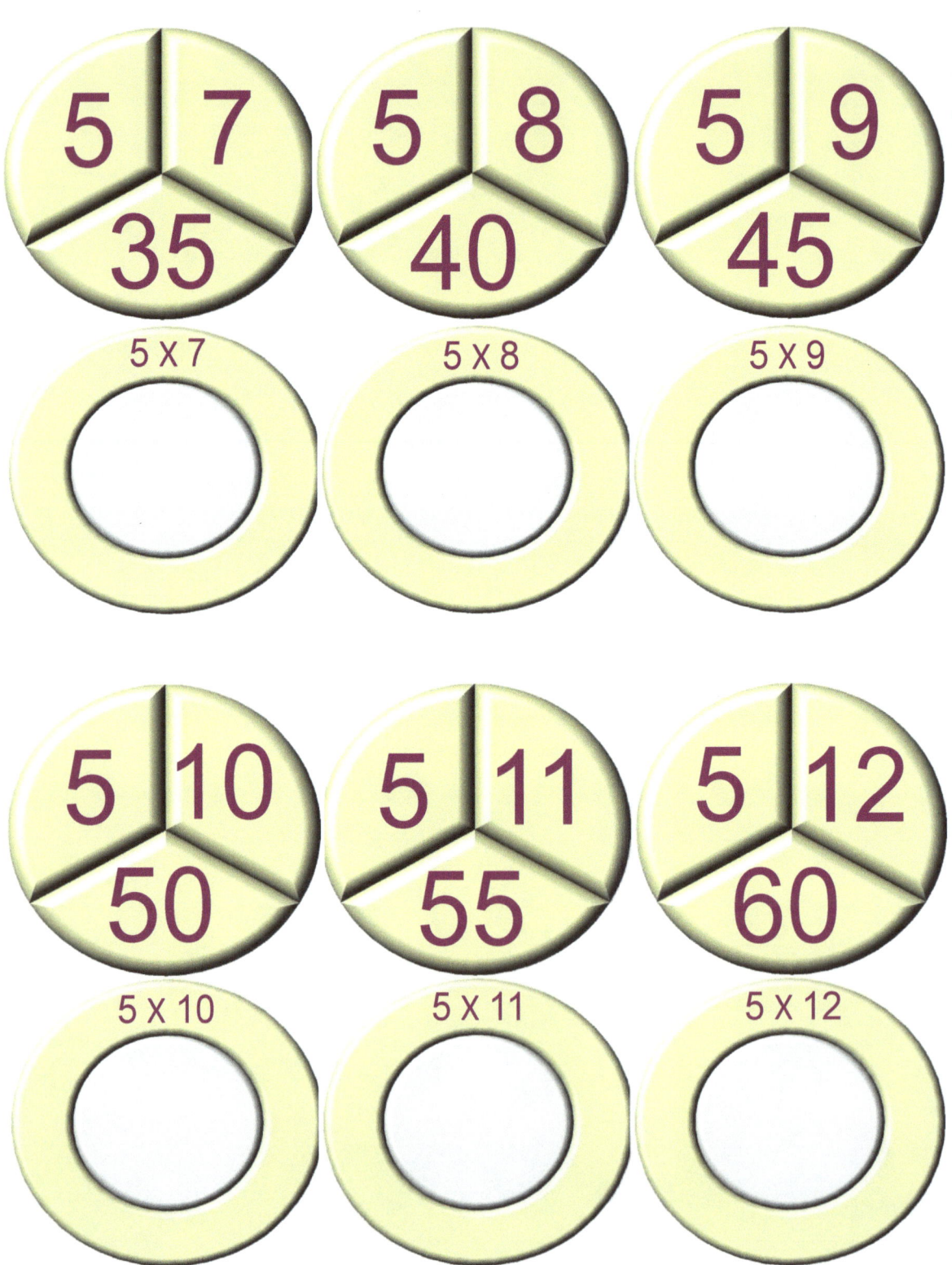

# Six times table

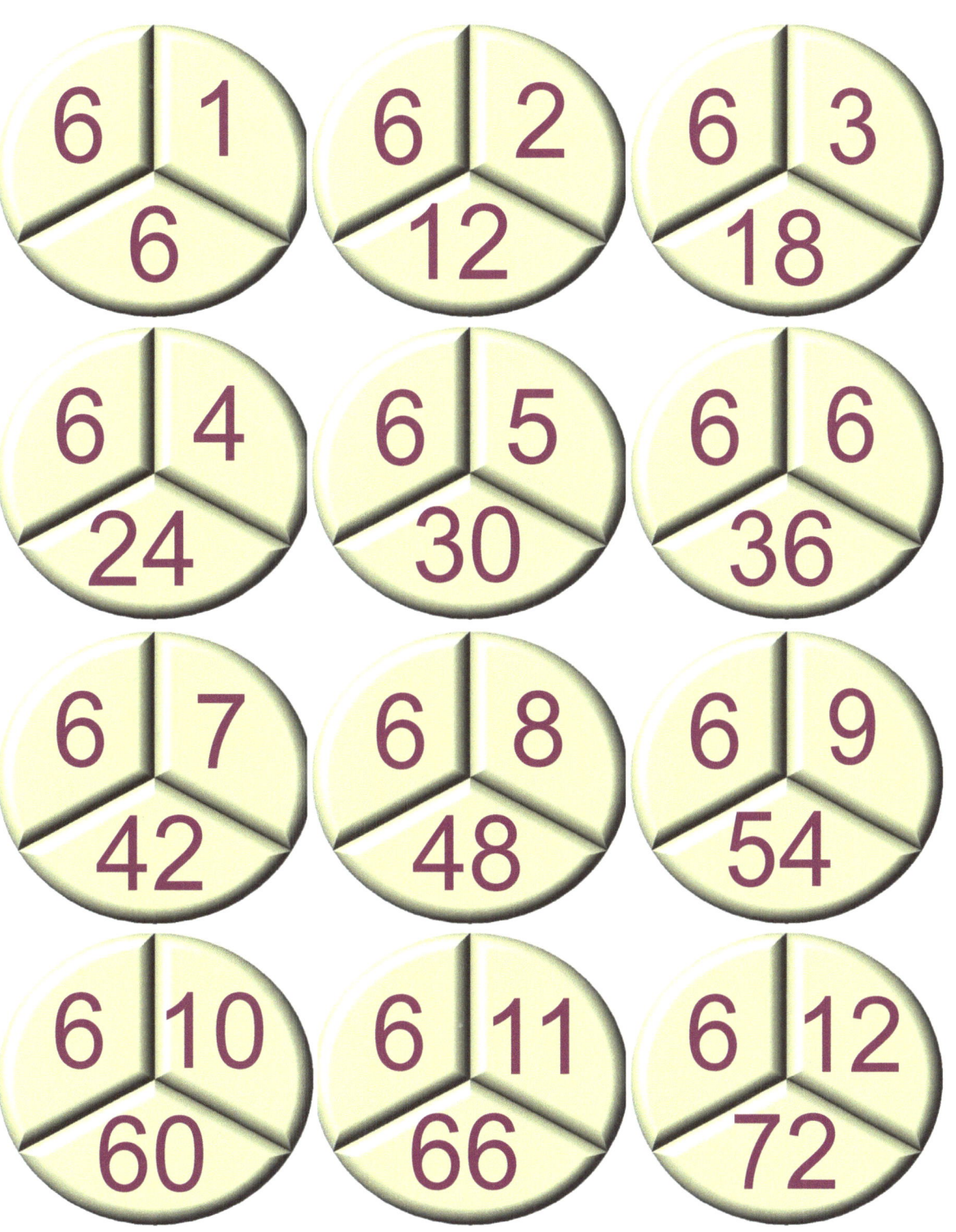

# Six times drills

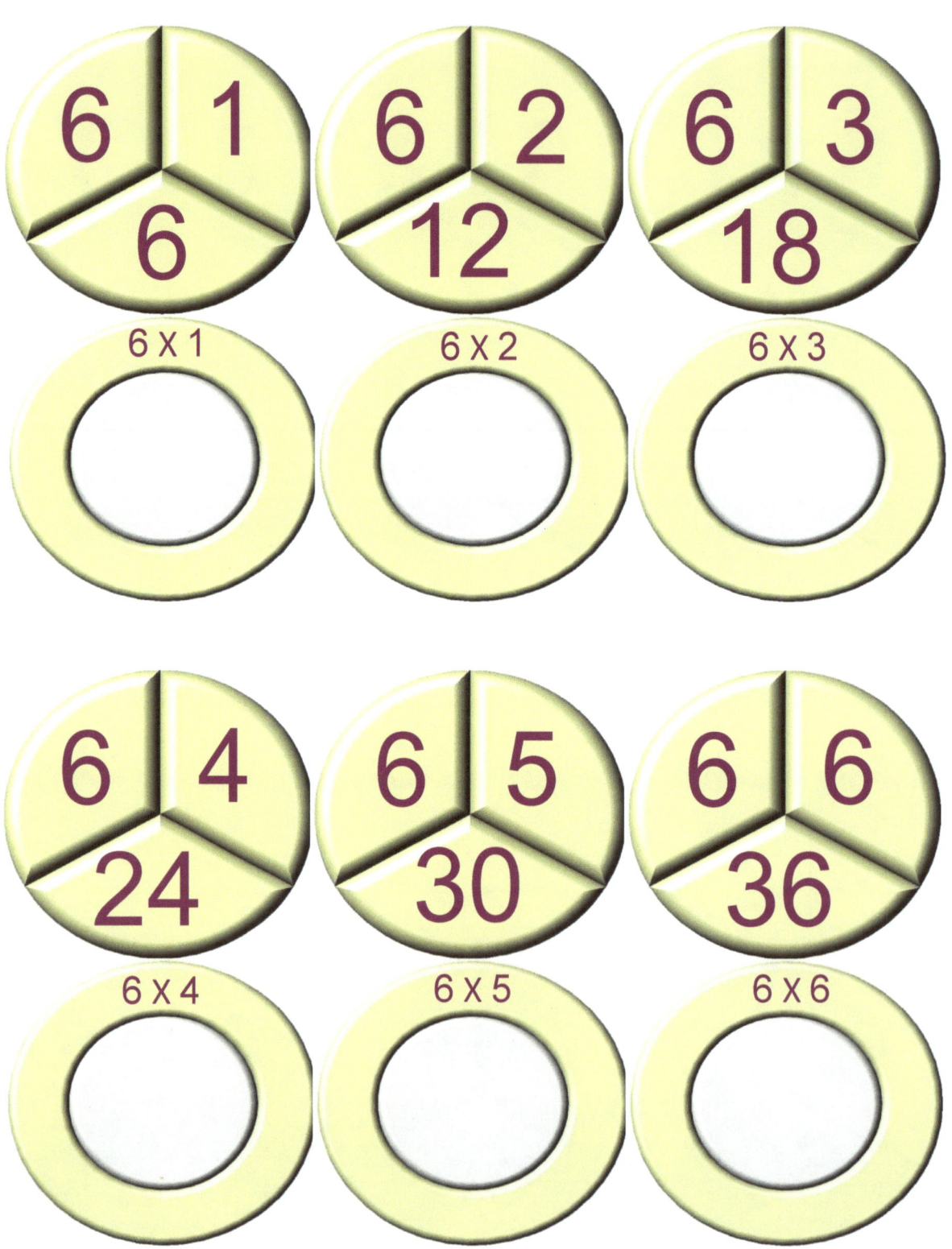

# Six times drills

# Seven times table

# Seven times drills

# Seven times drills

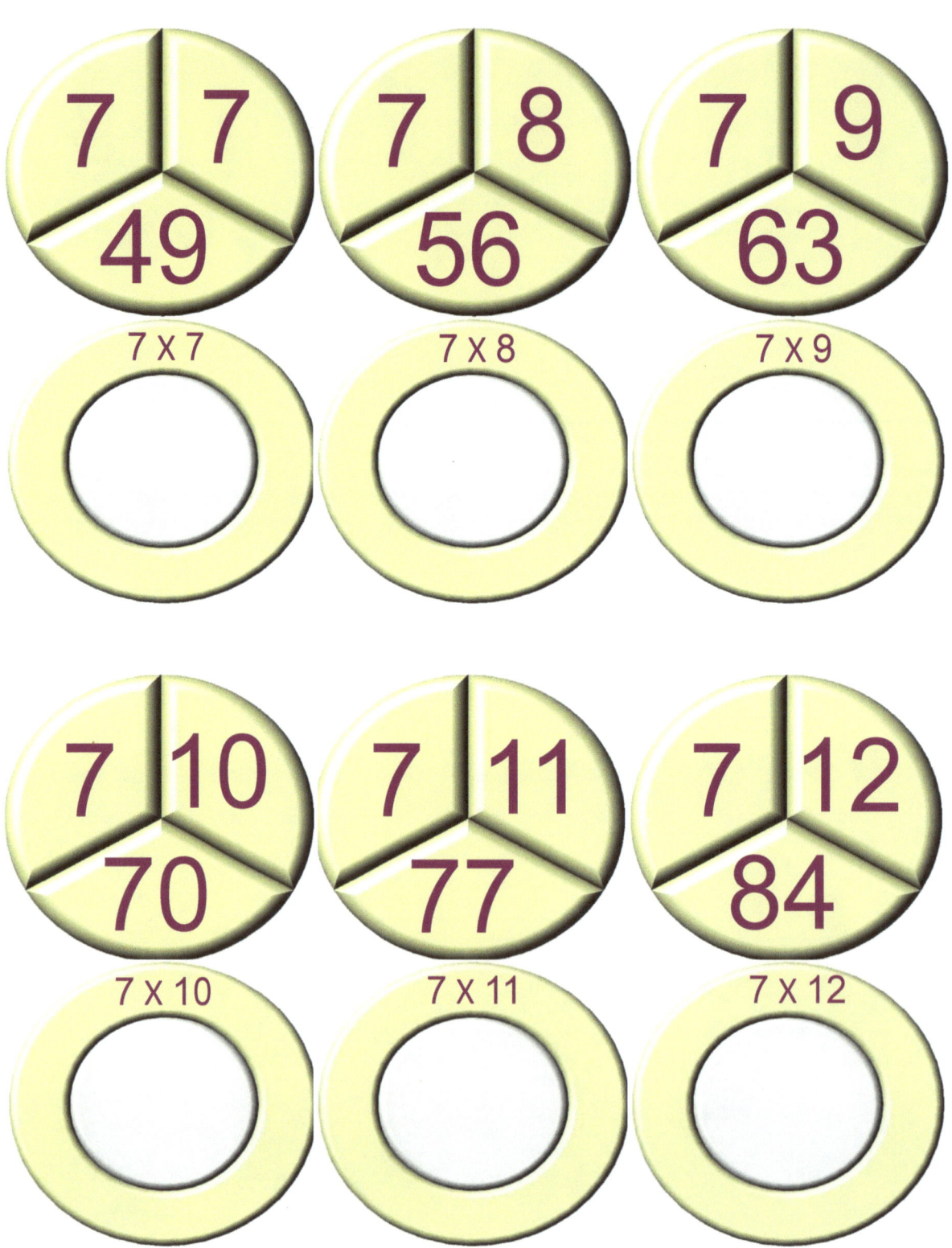

# Eight times table

# Eight times drills

# Eight times drills

# Nine times table

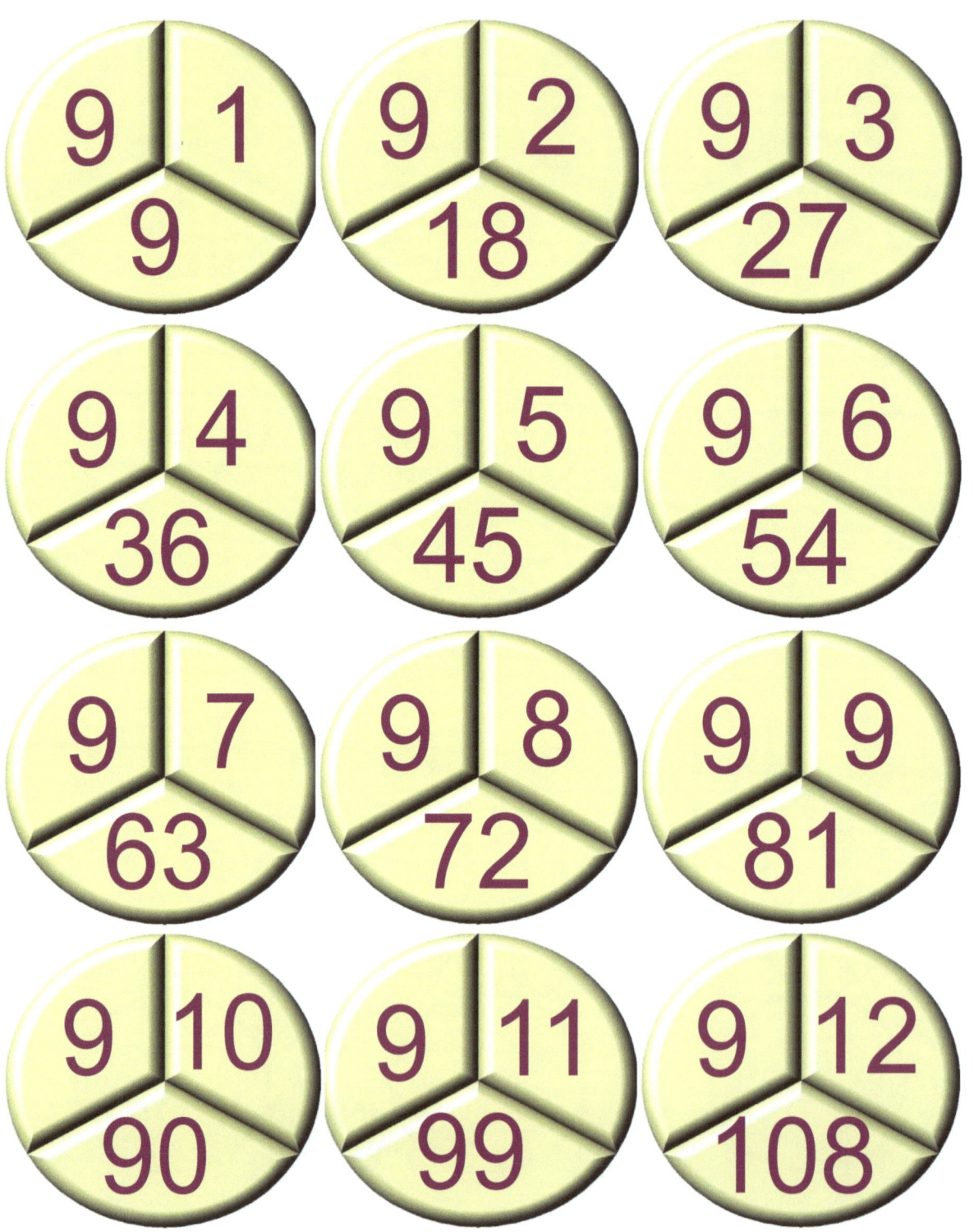

# Nine times drills

# Nine times drills

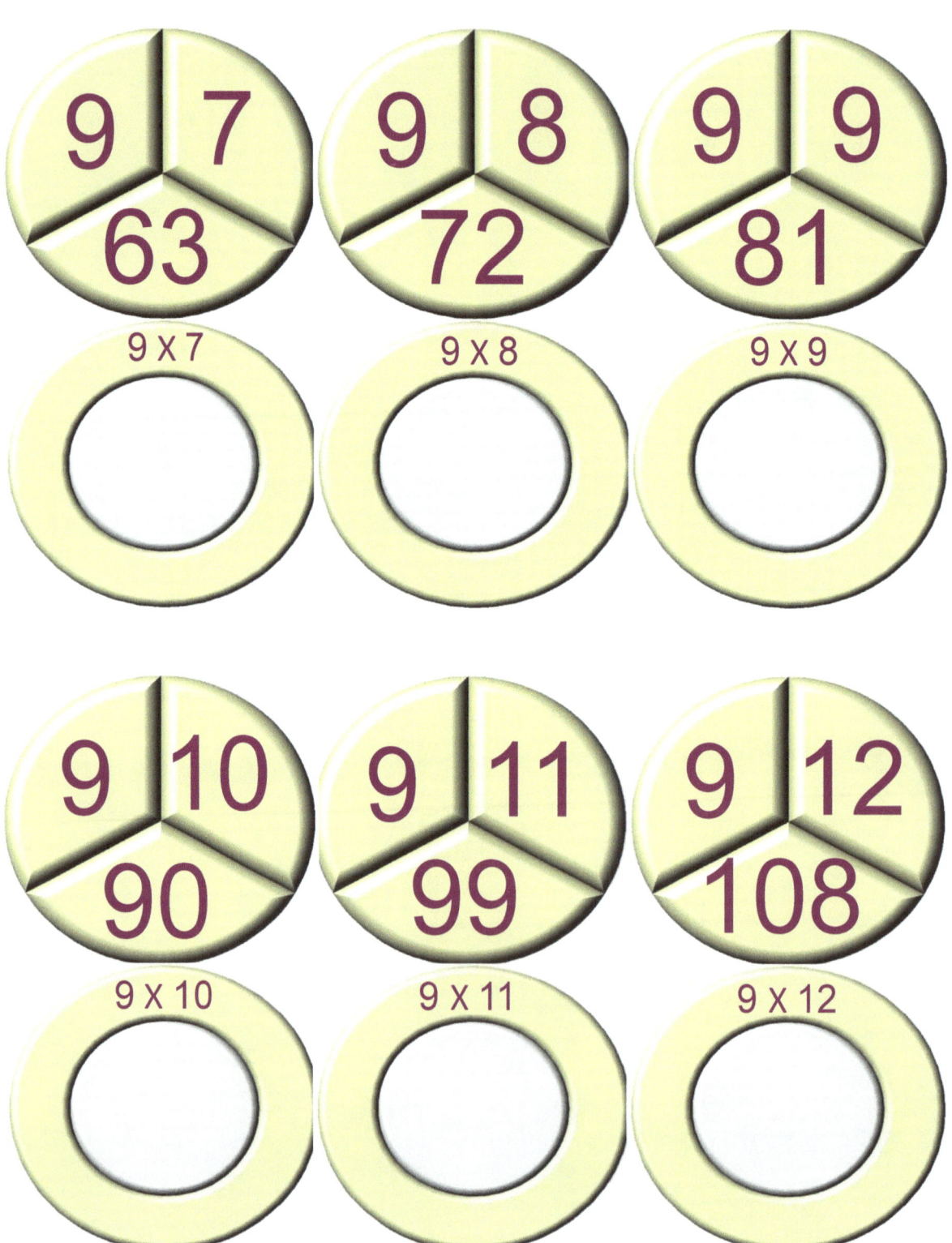

# Ten times table

# Ten times drills

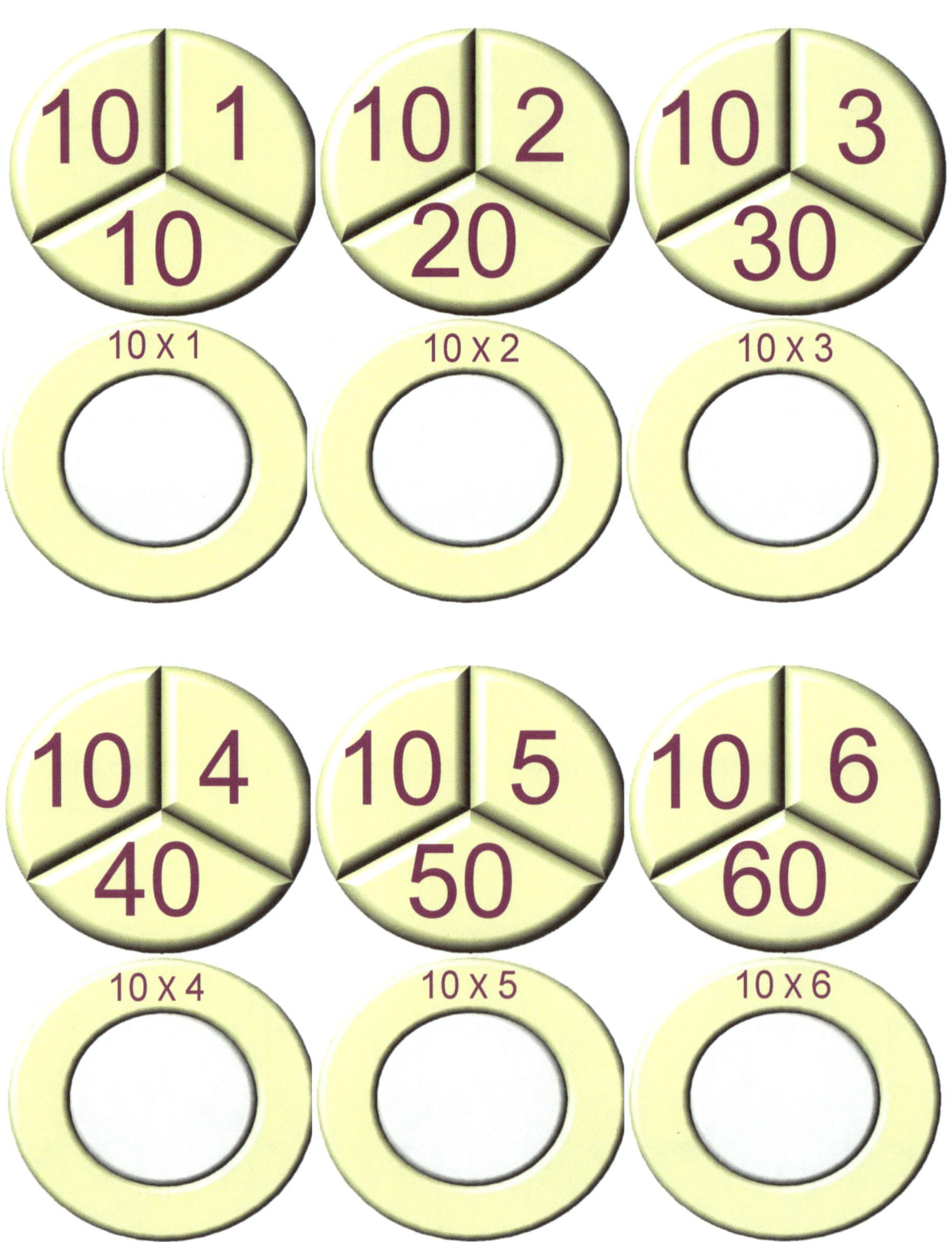

# Ten times drills

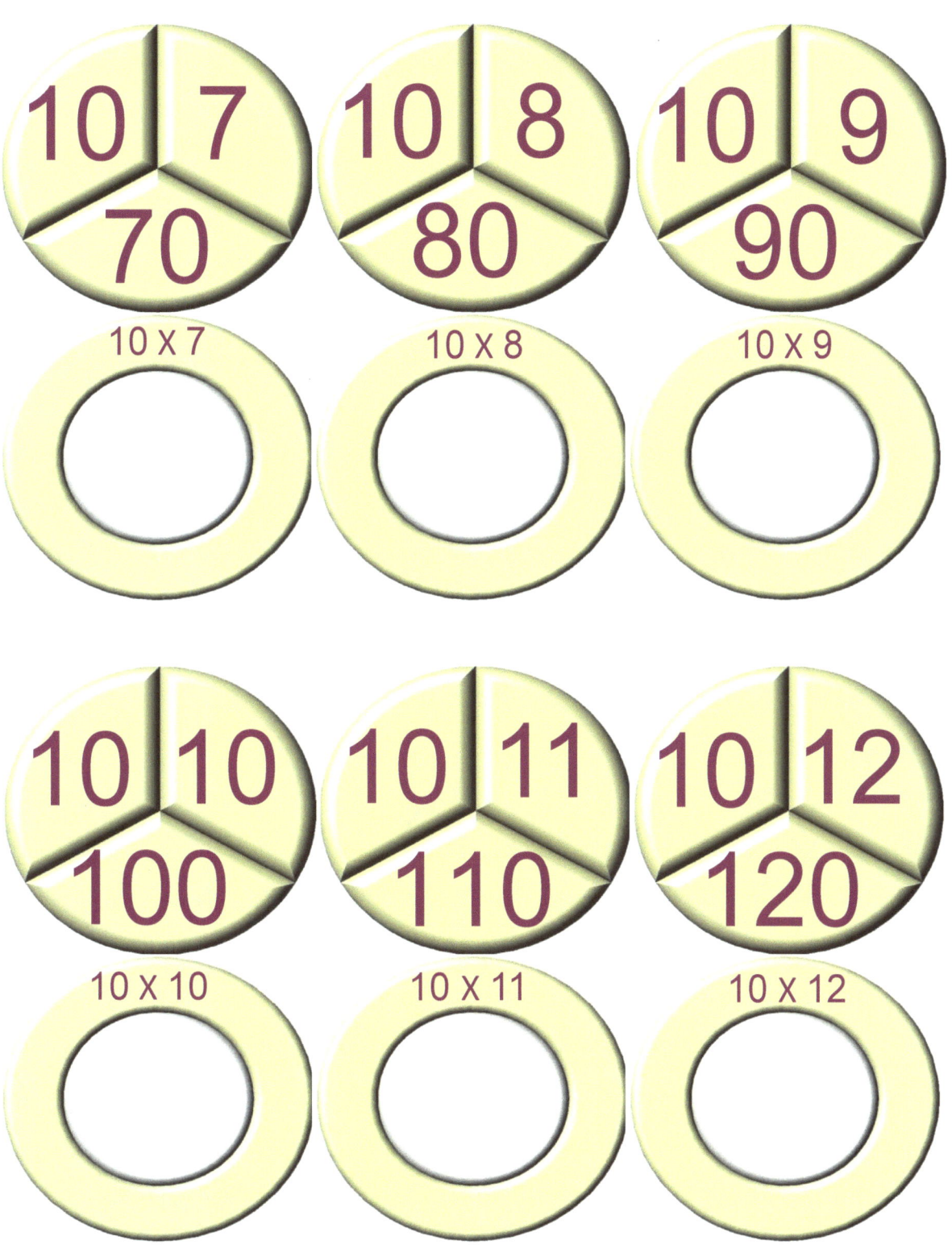

# Eleven times table

# Eleven times drills

# Eleven times drills

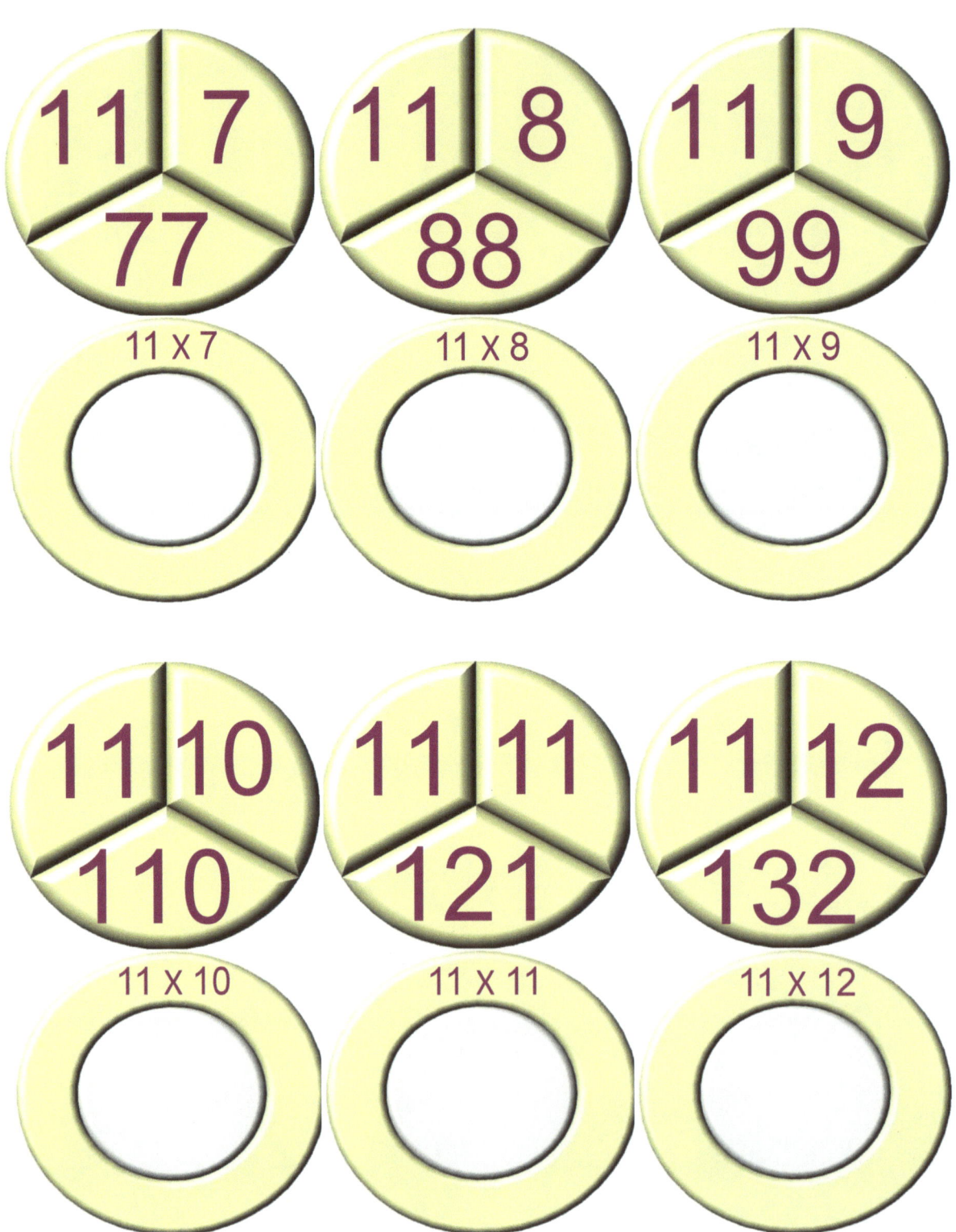

# Twelve times table

# Twelve times drills

# Twelve times drills

# One times drills

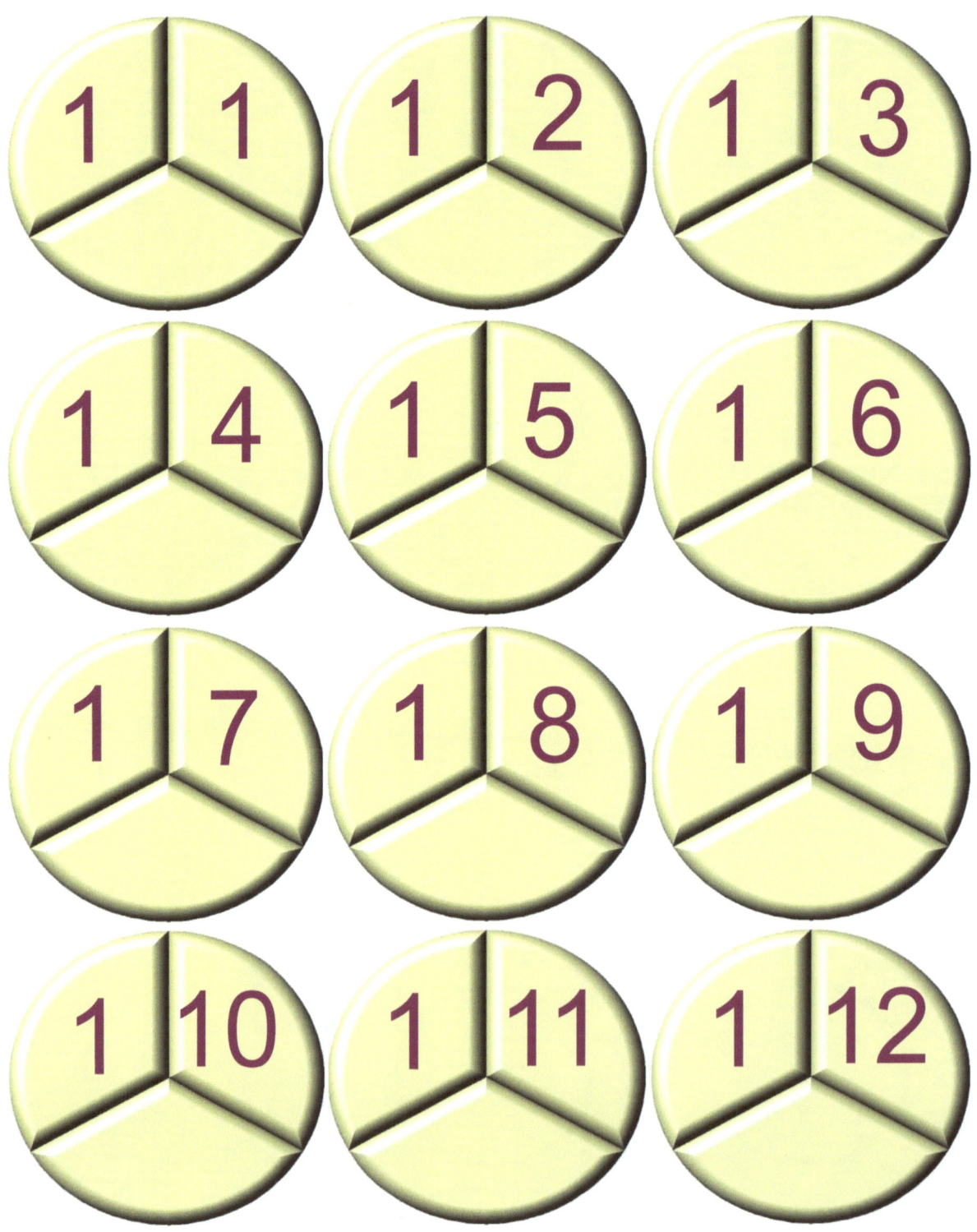

# Two times drills

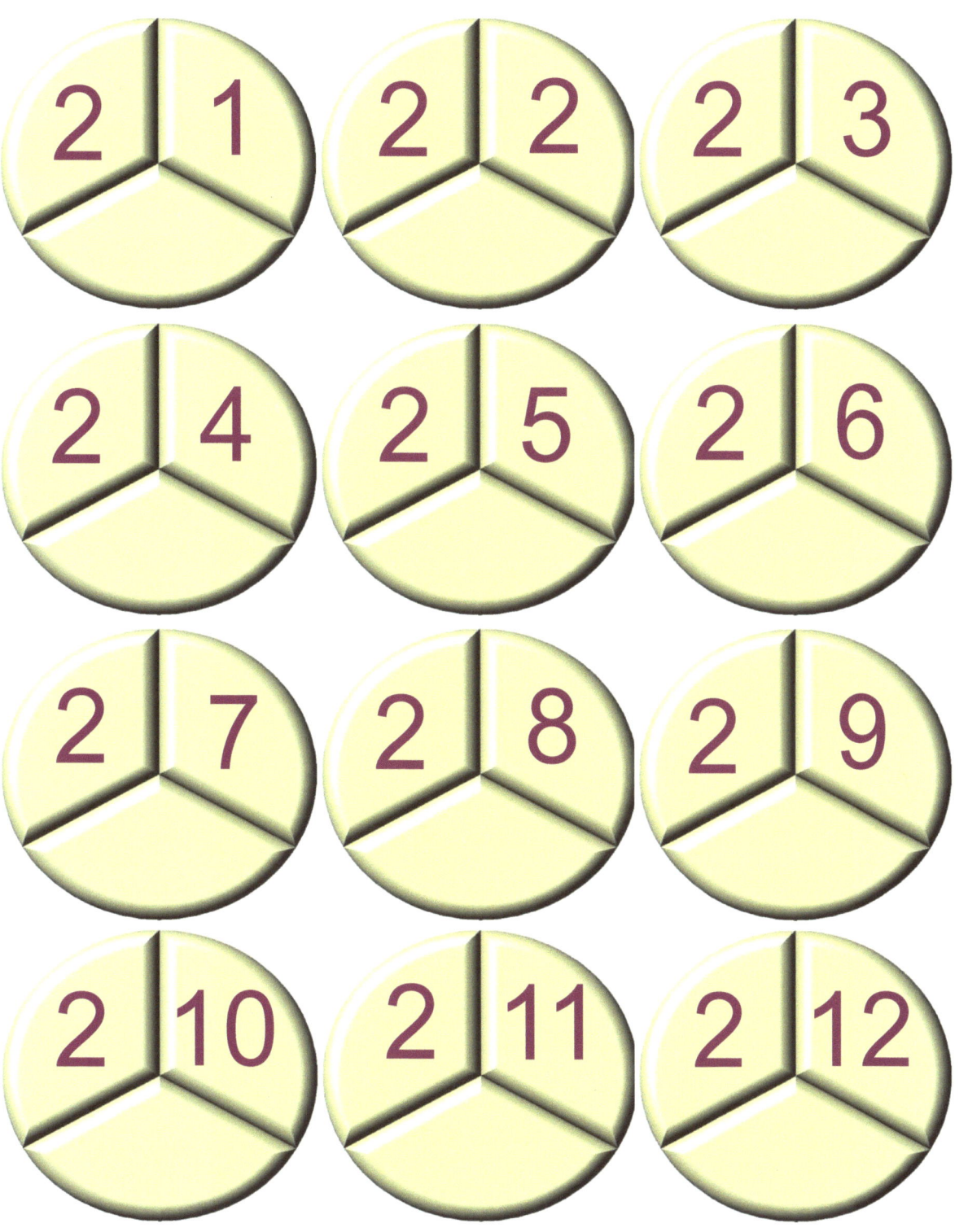

# Three times drills

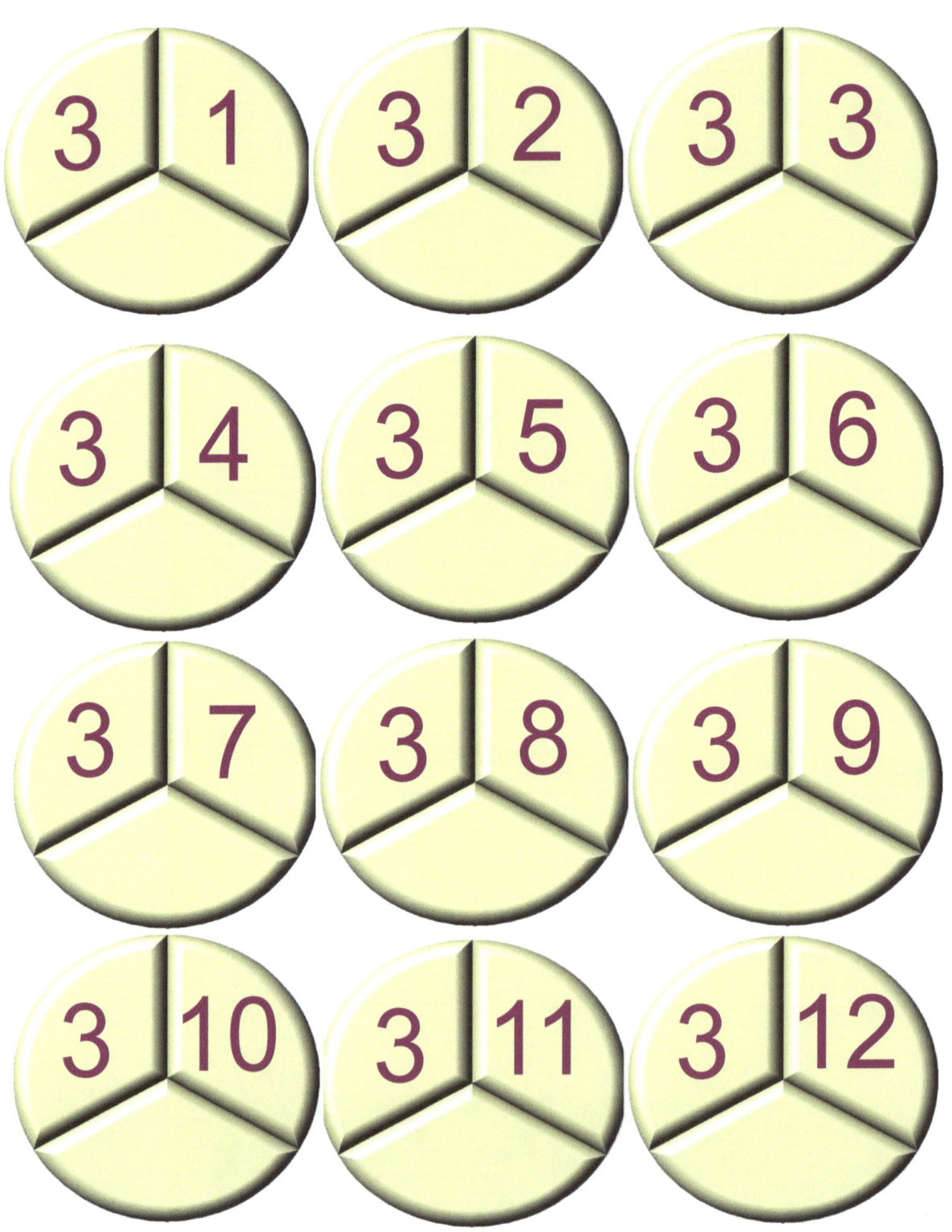

# Four times drills

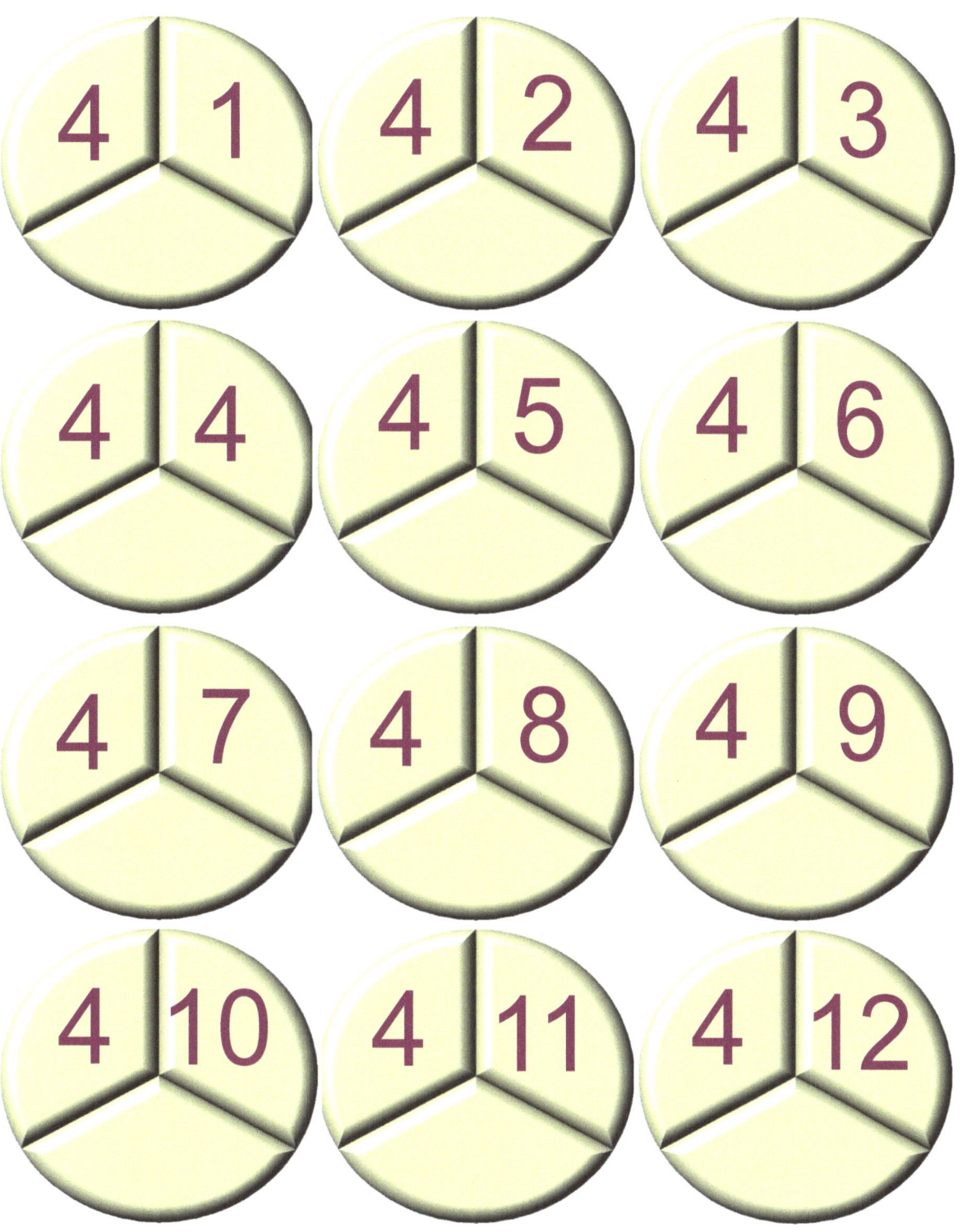

# Five times drills

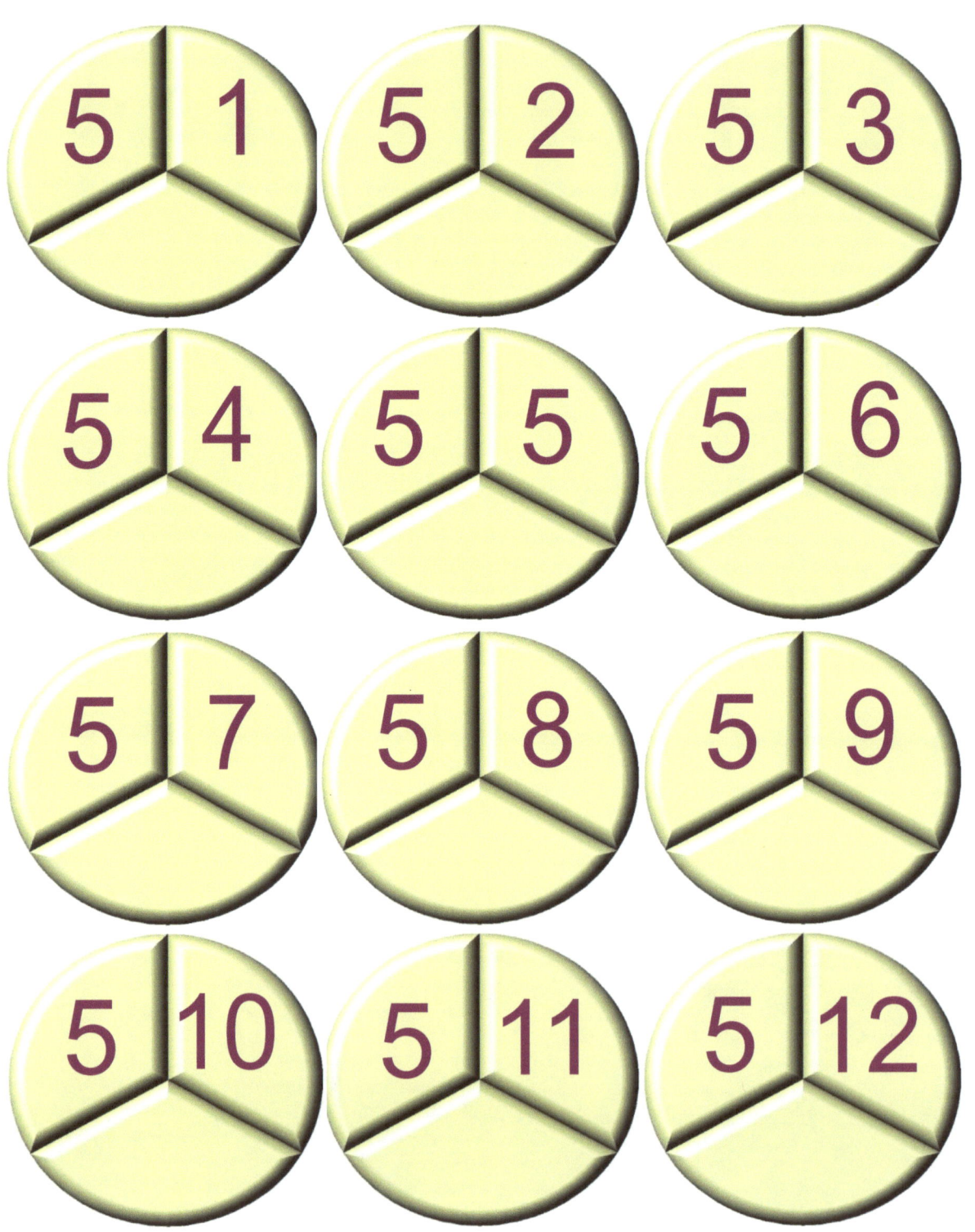

# Six times drills

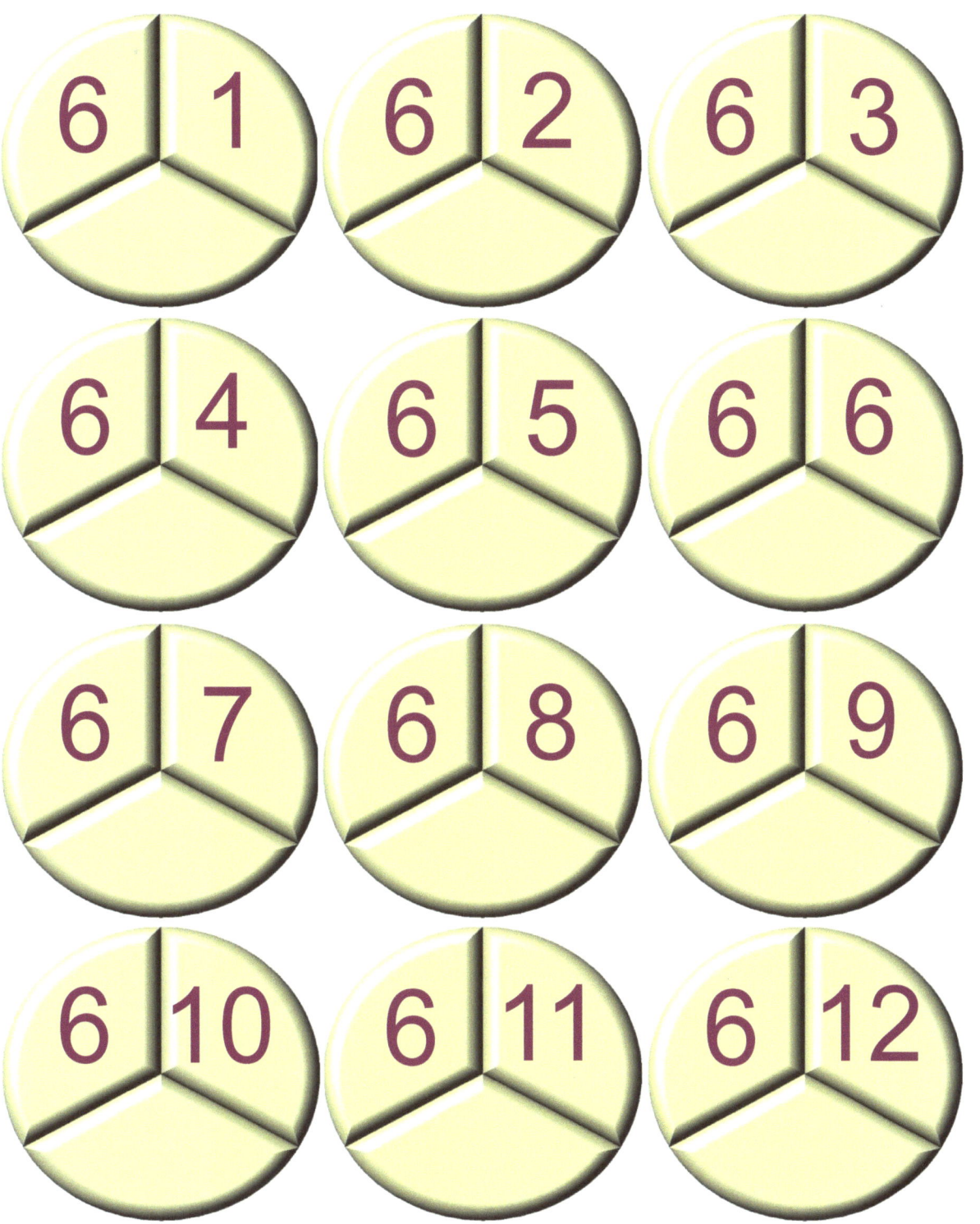

# Seven times drills

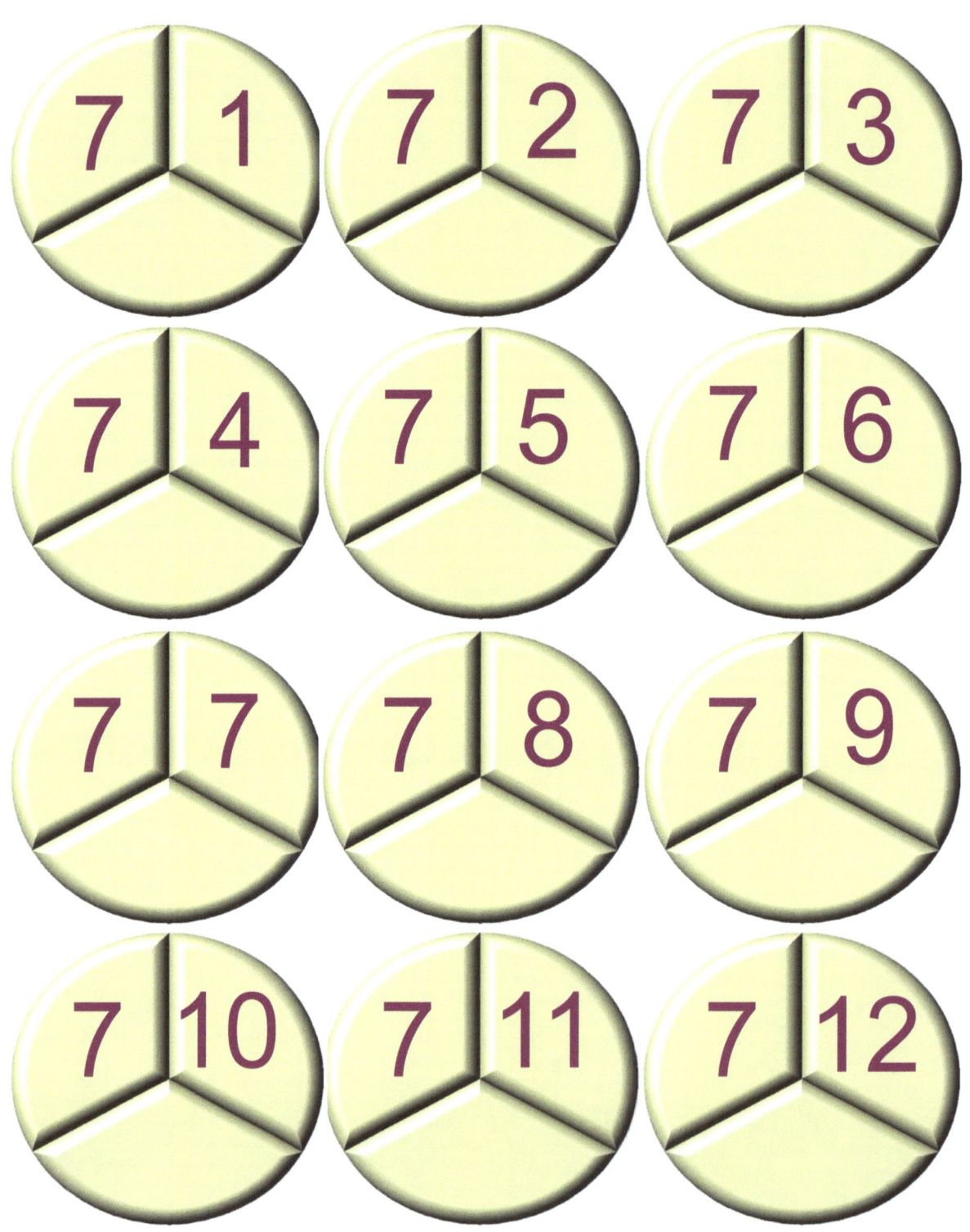

# Eight times drills

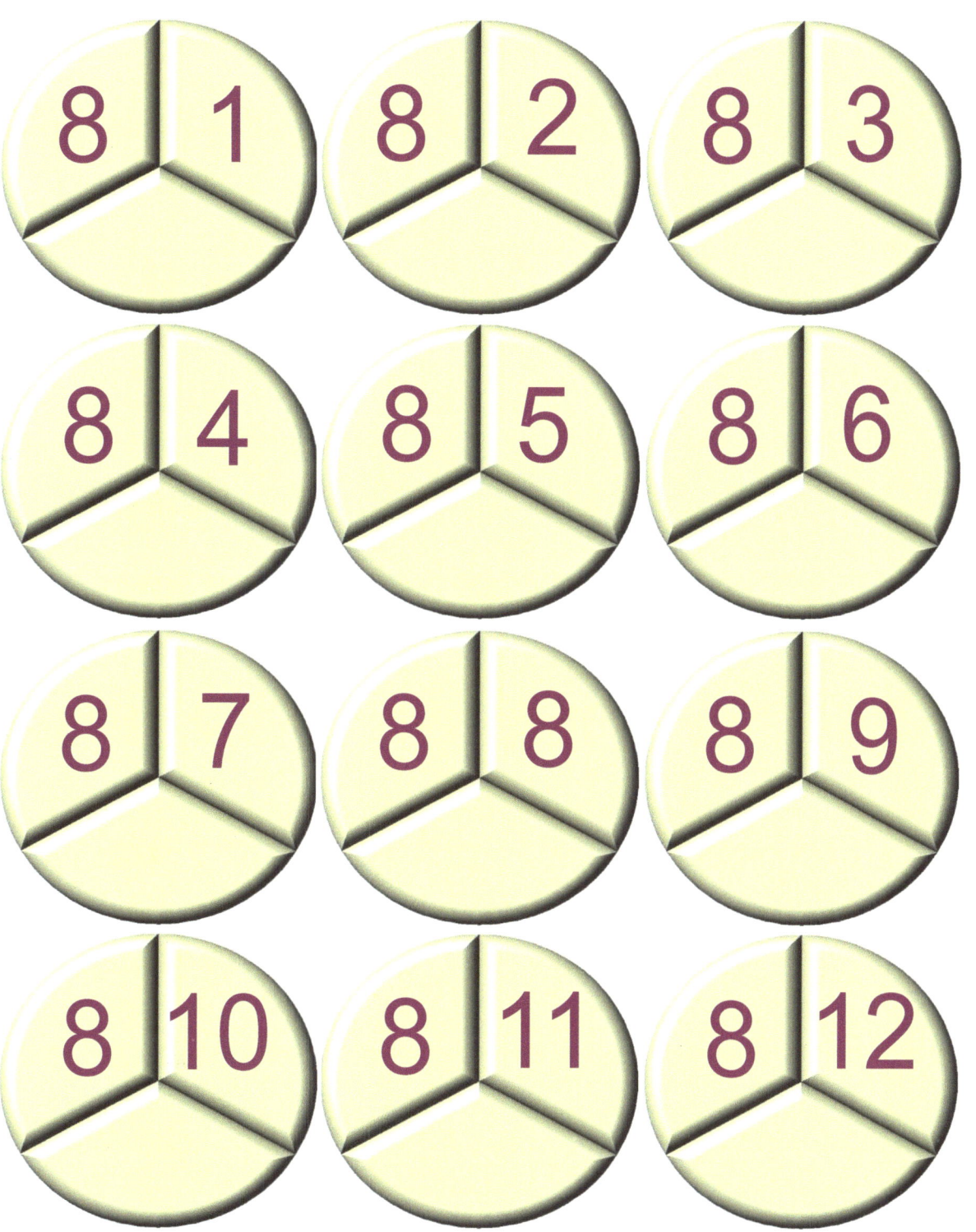

# Nine times drills

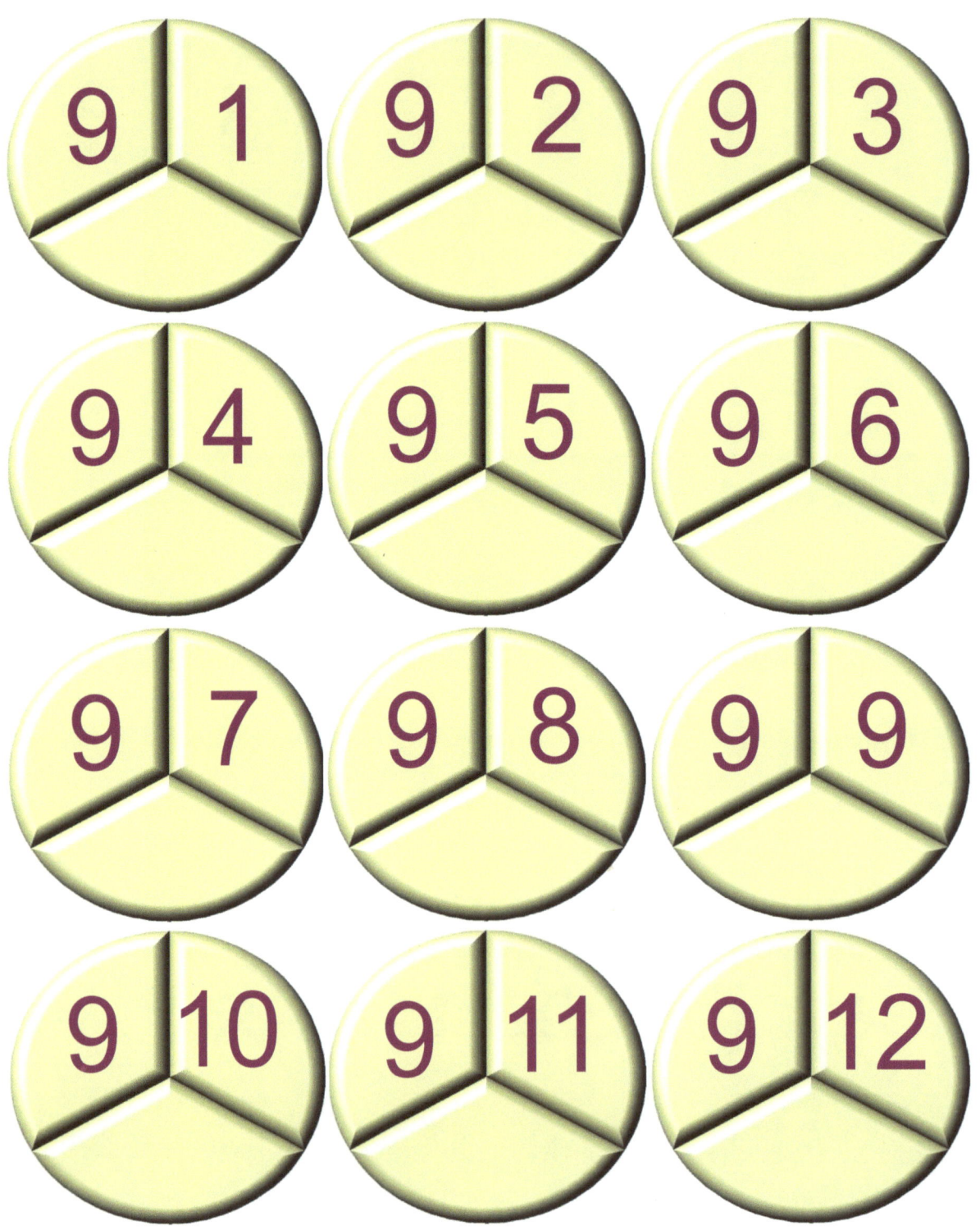

# Ten times drills

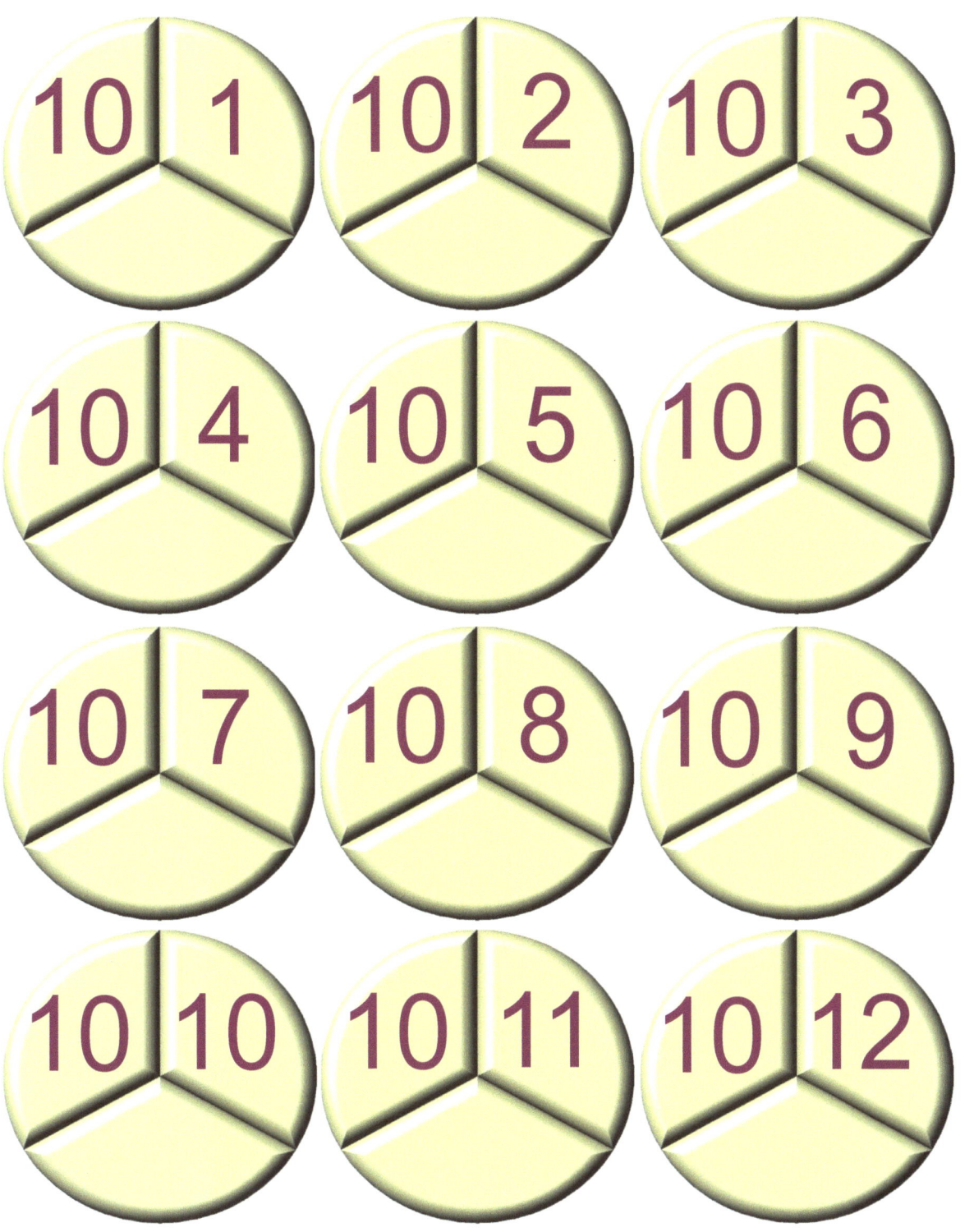

# Eleven times drills

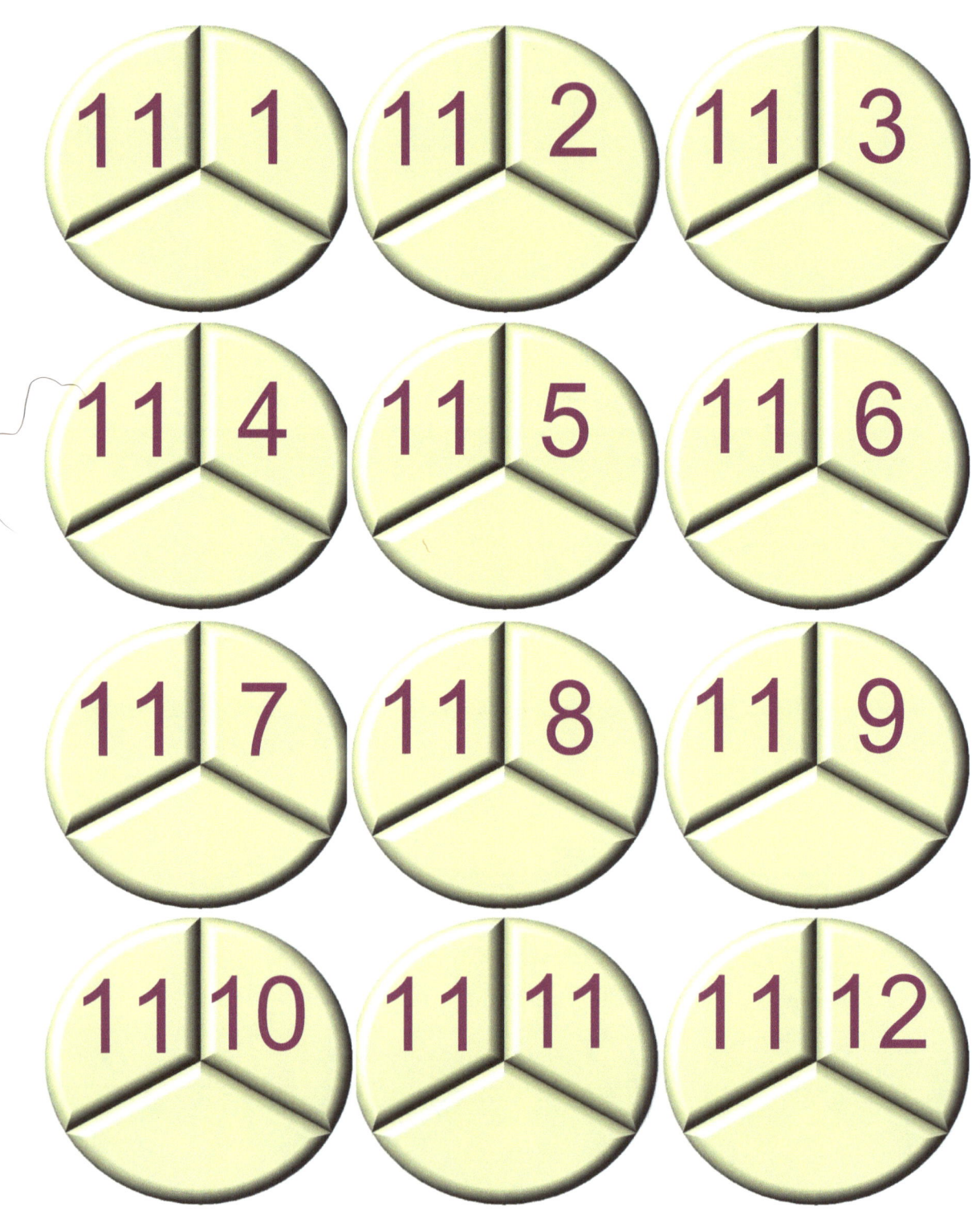

# Twelve times drills

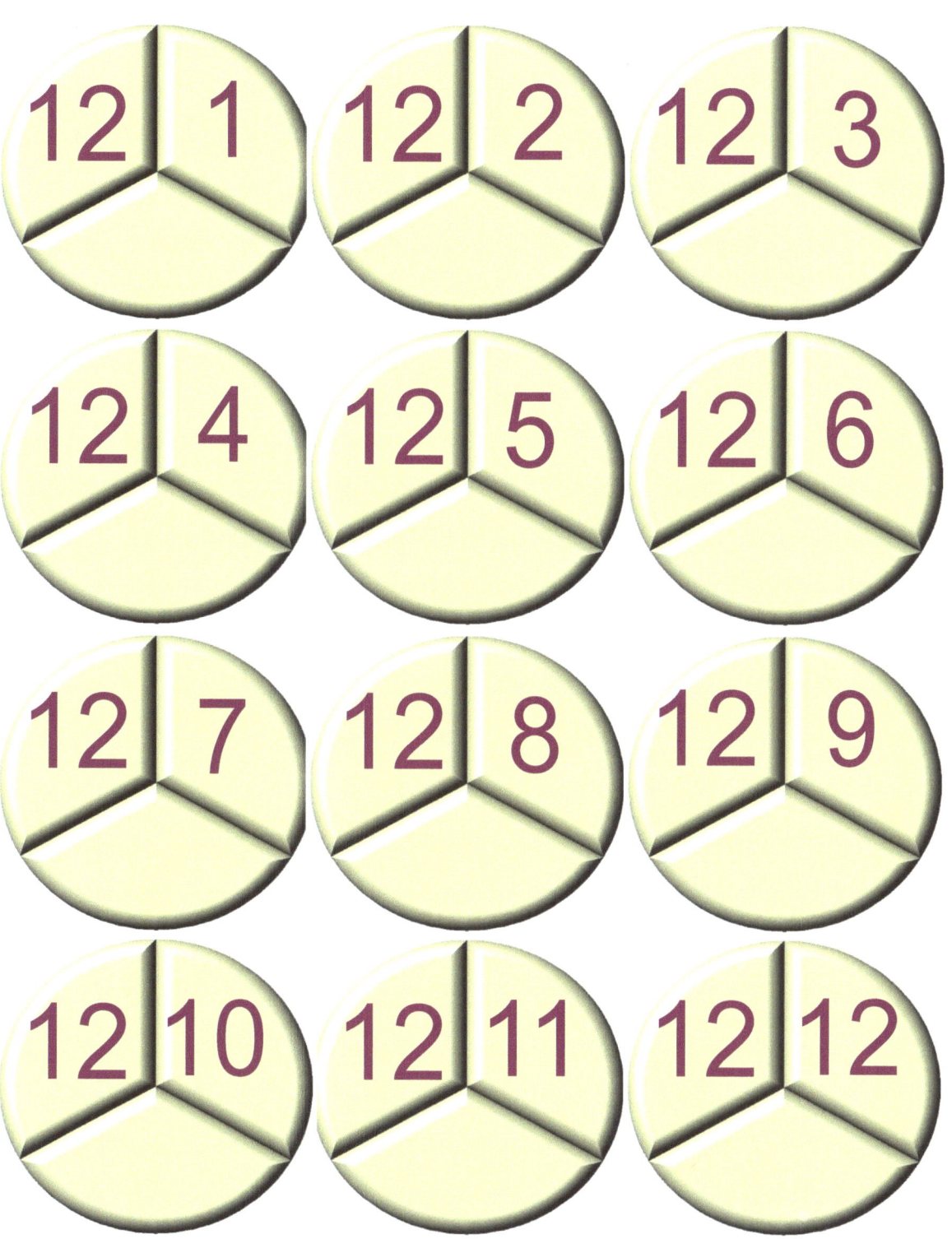

# One times drills

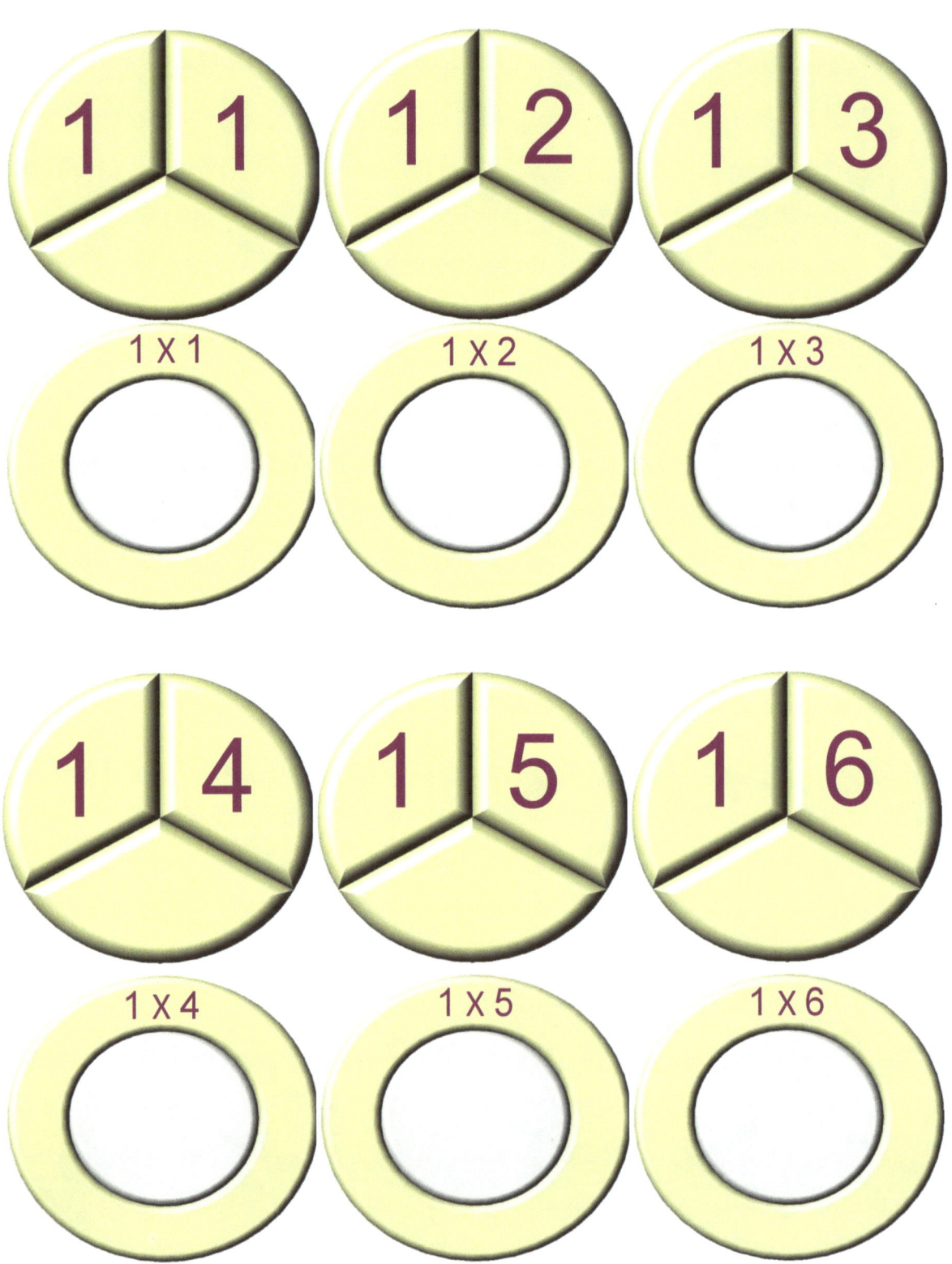

# One times drills

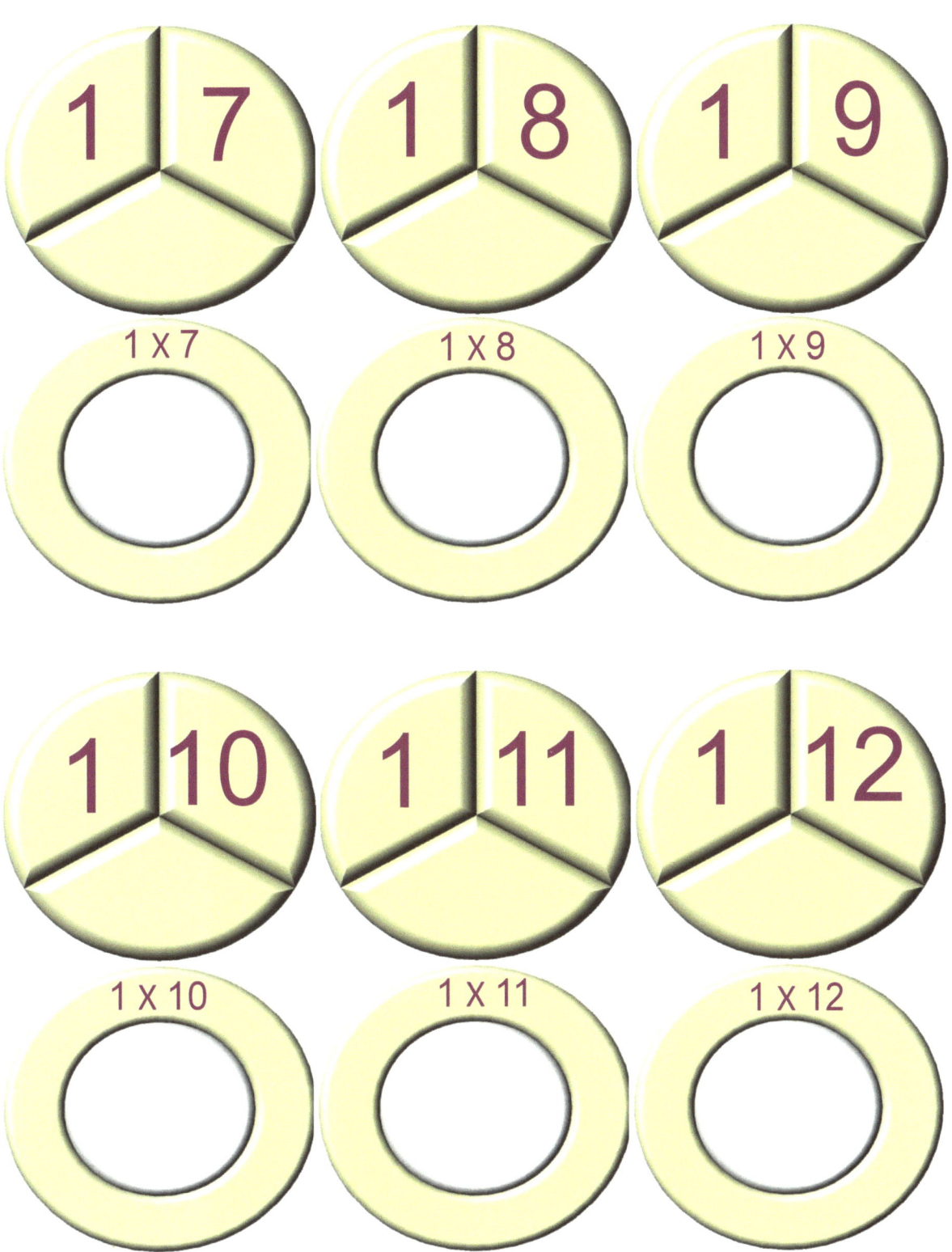

# Two times drills

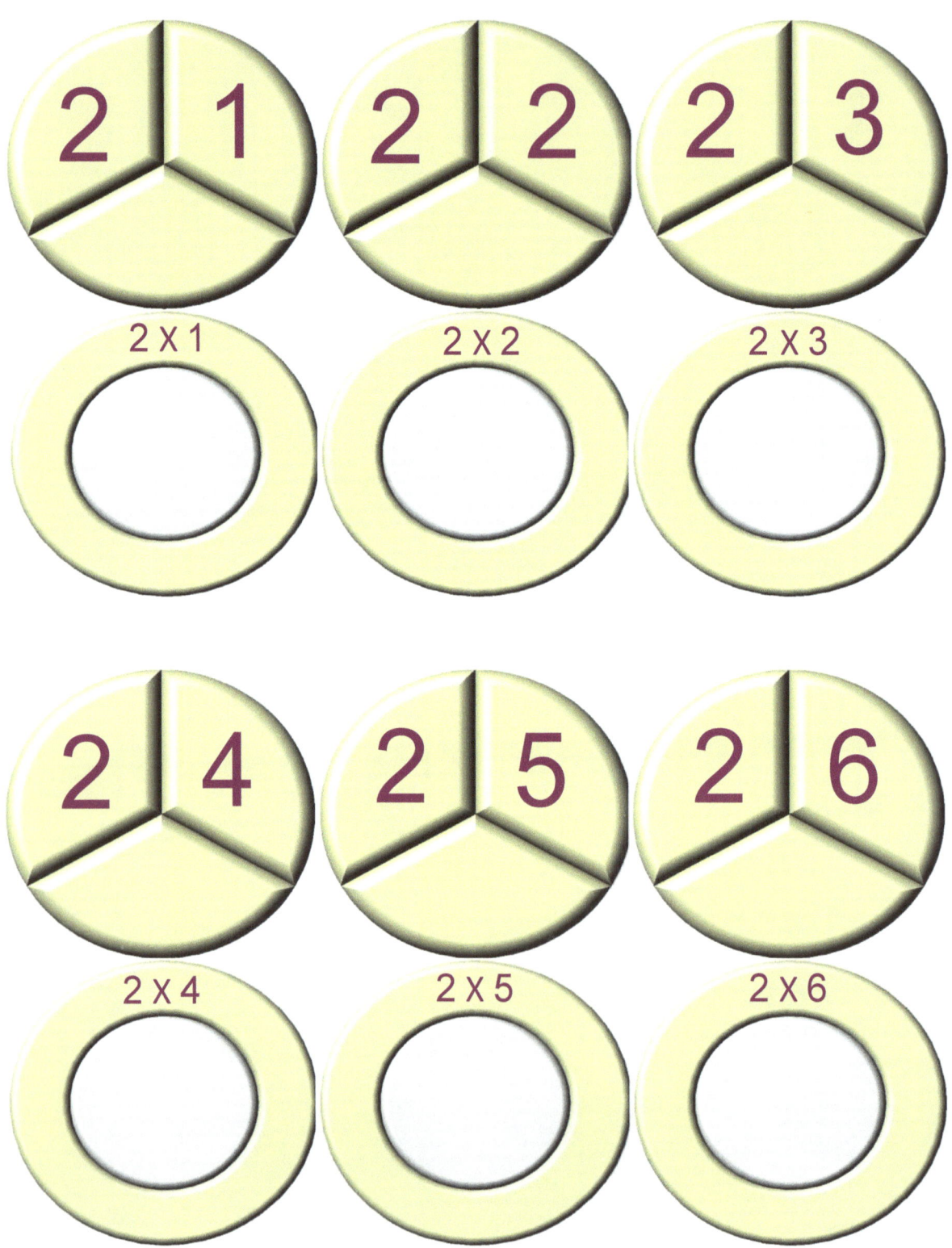

# Two times drills

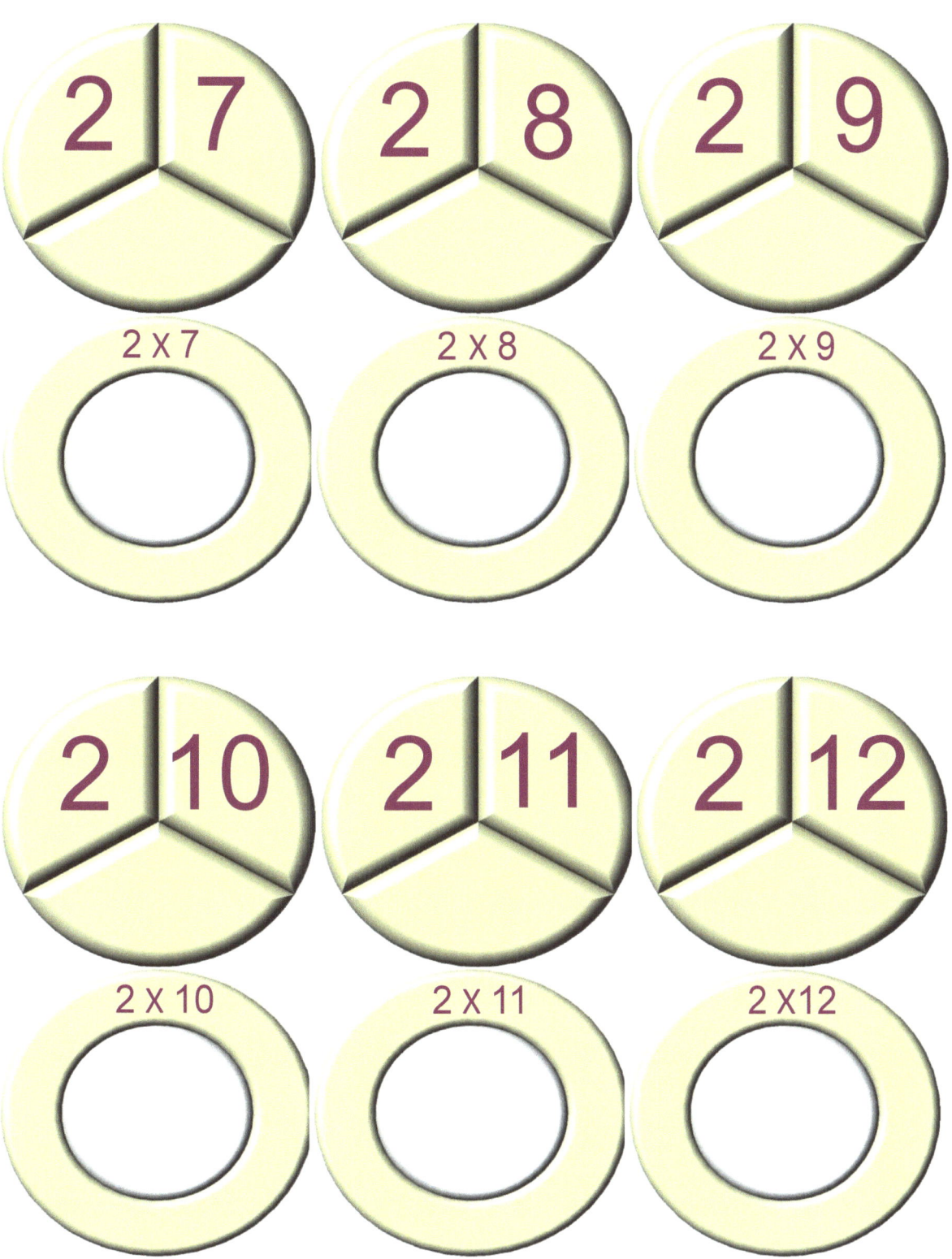

# Three times drills

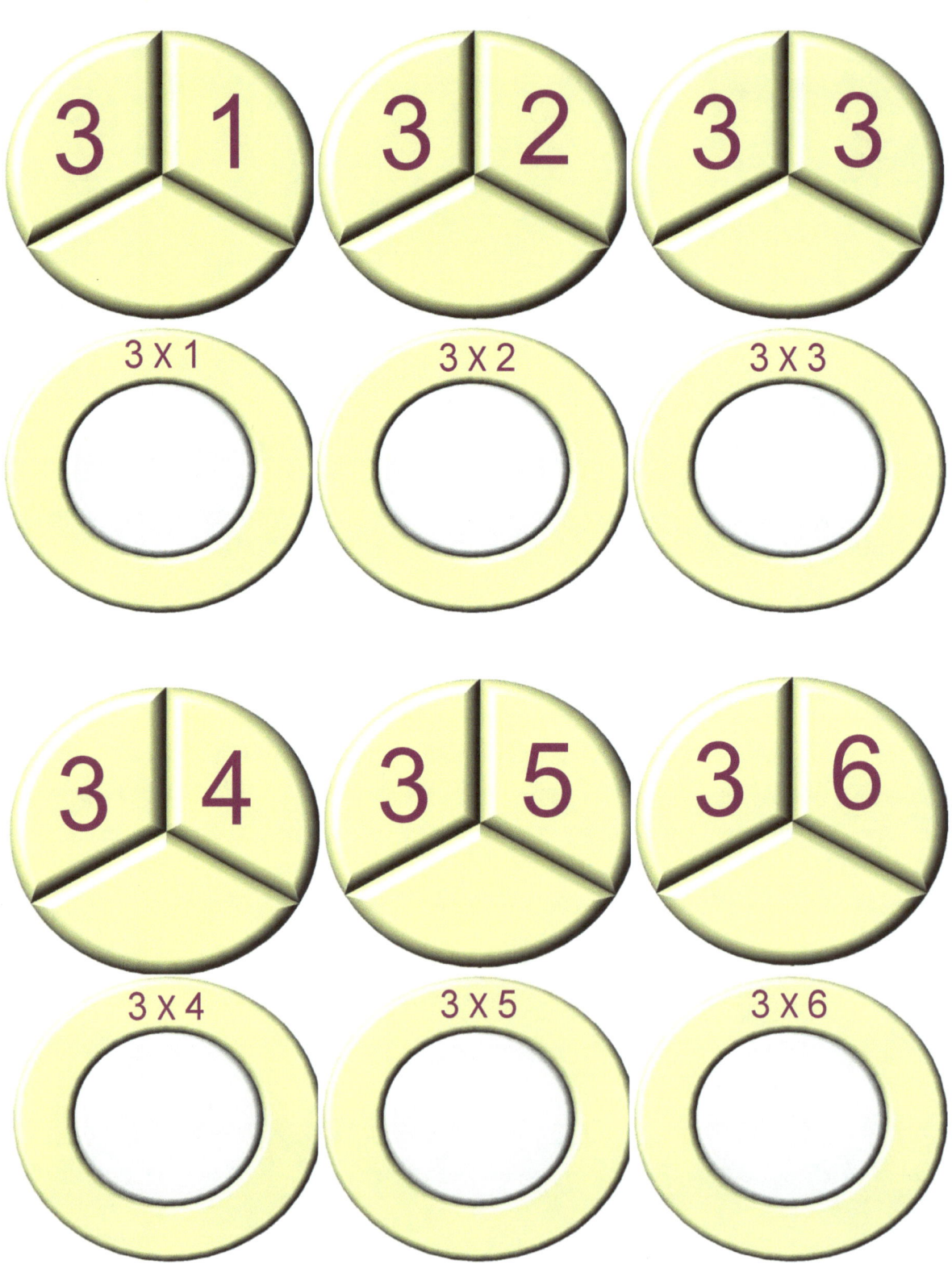

# Three times drills

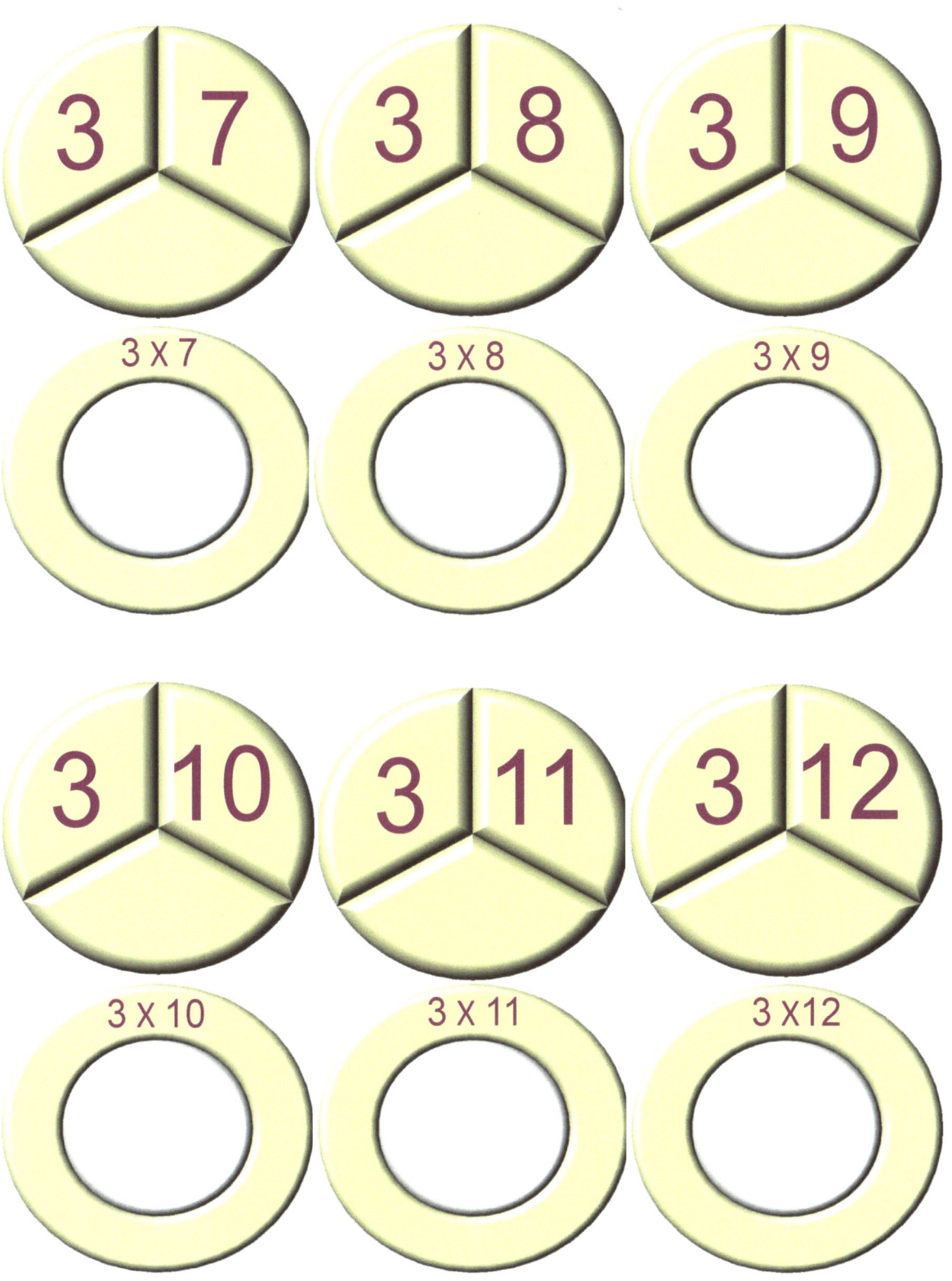

# Four times drills

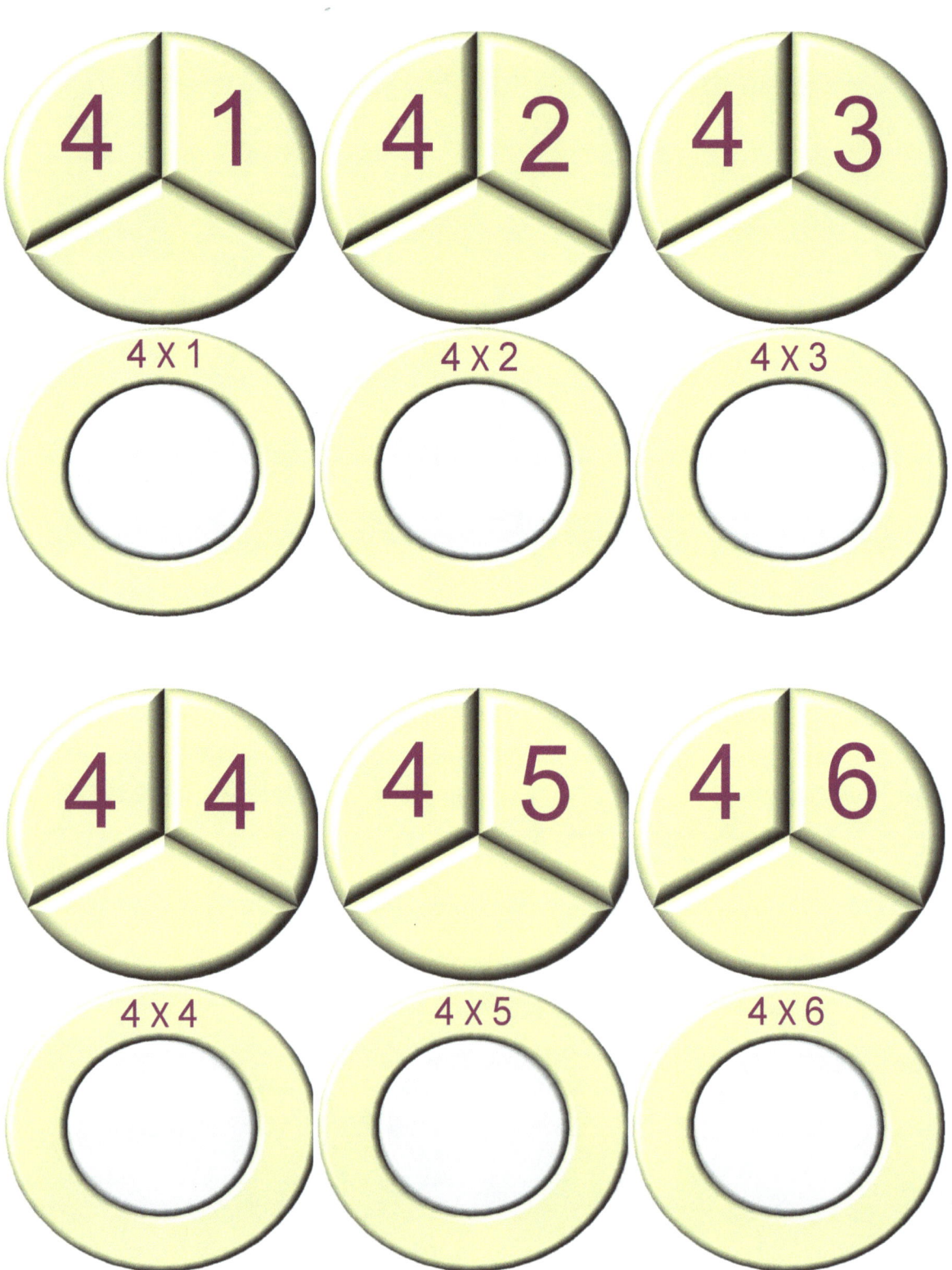

# Four times drills

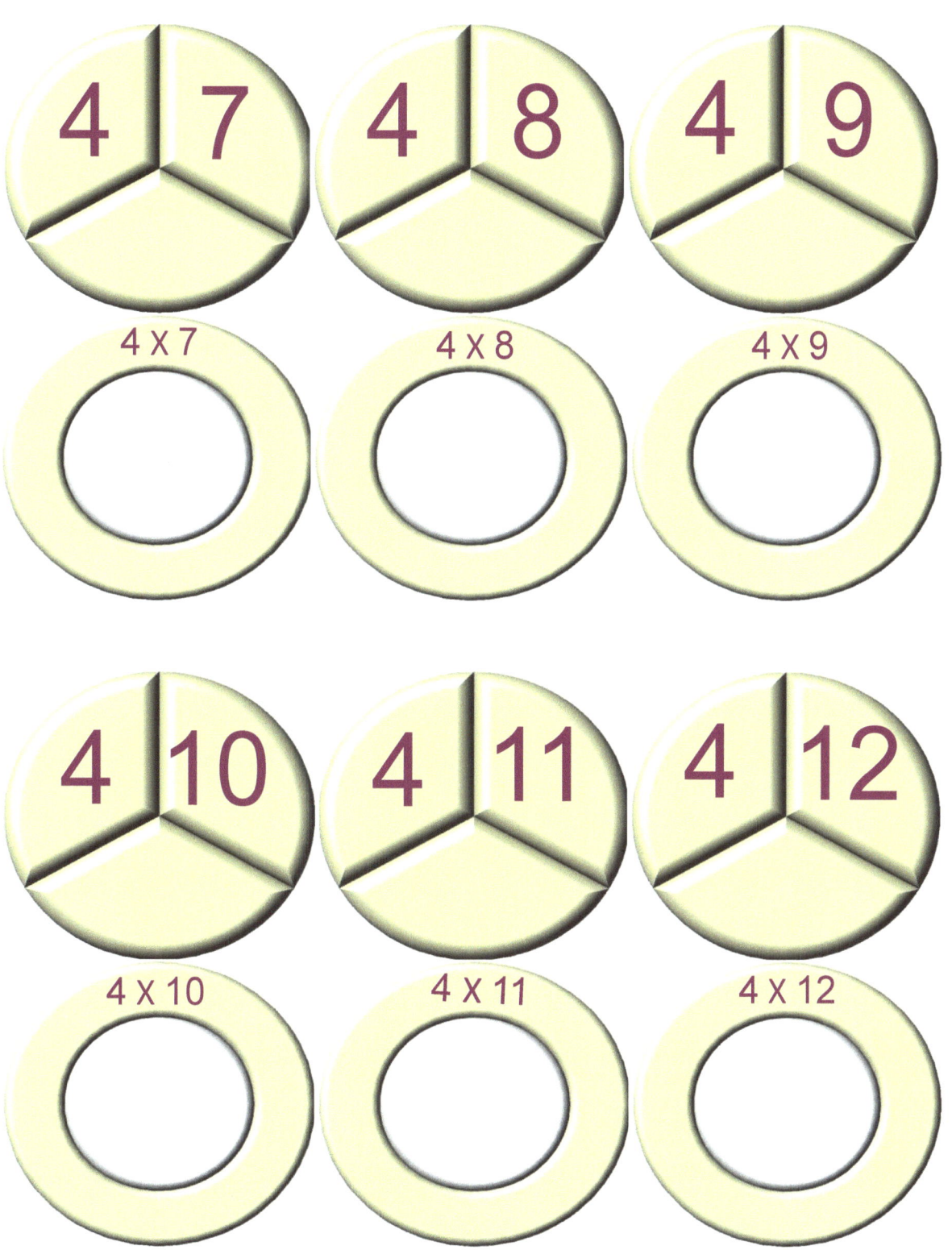

# Five times drills

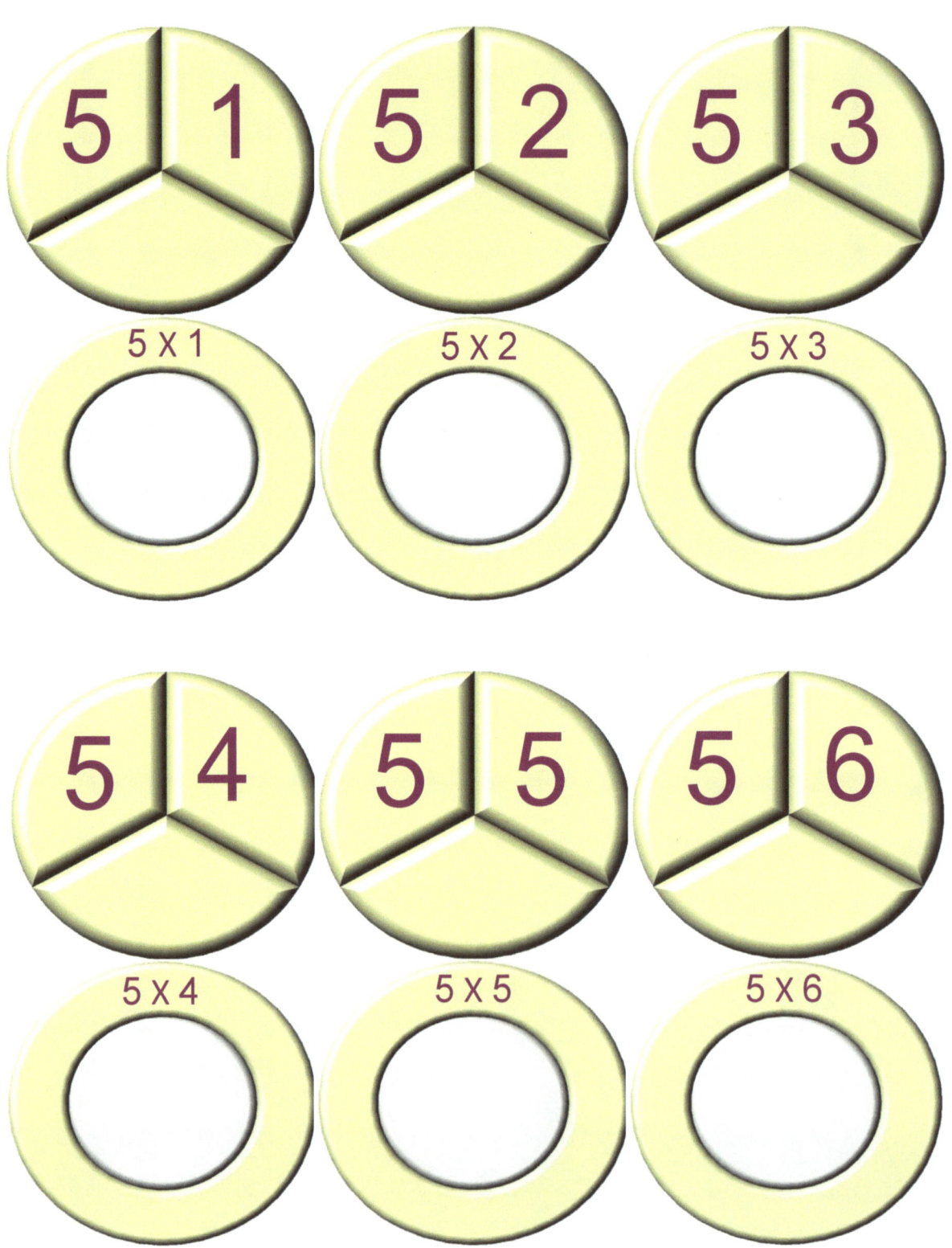

# Five times drills

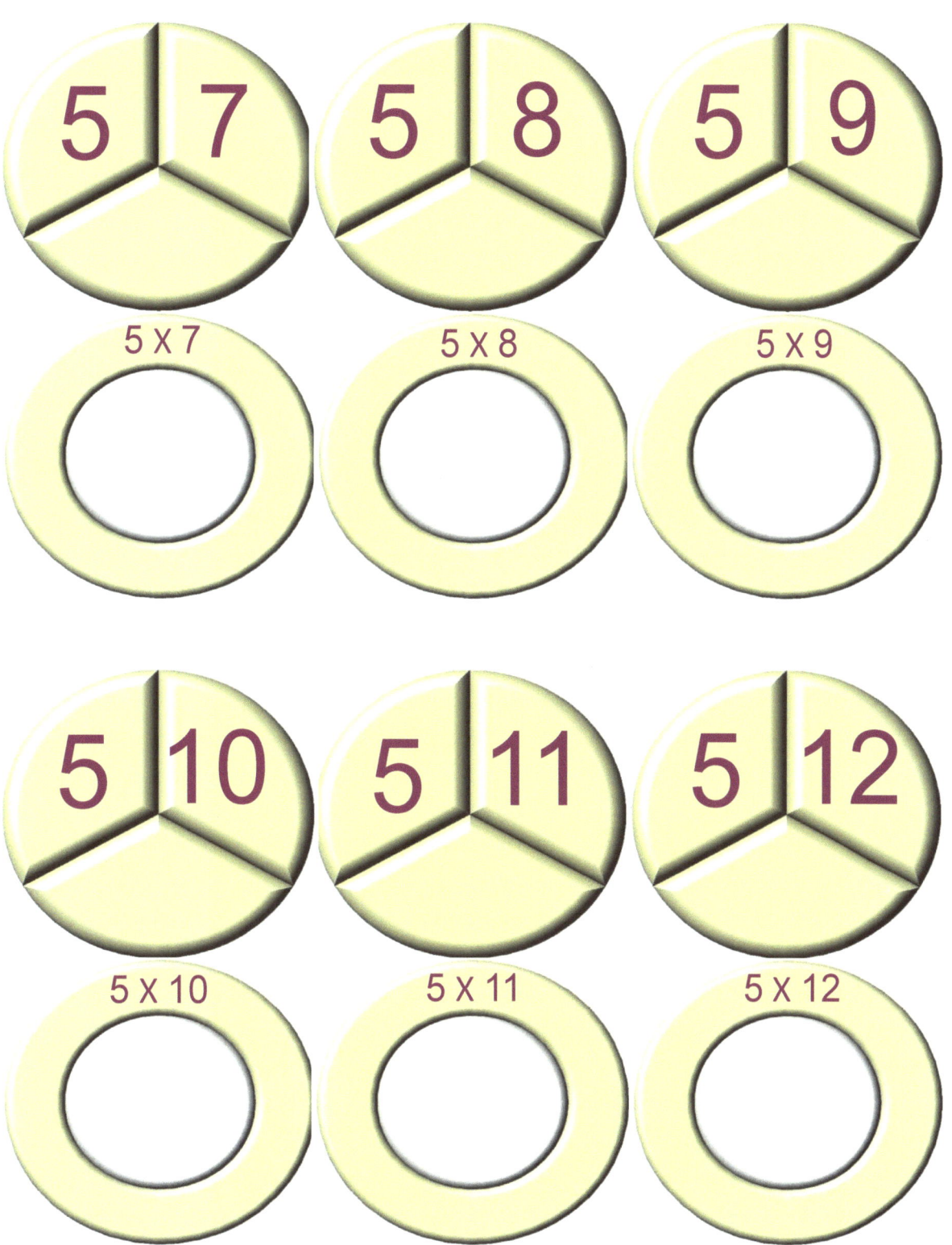

# Six times drills

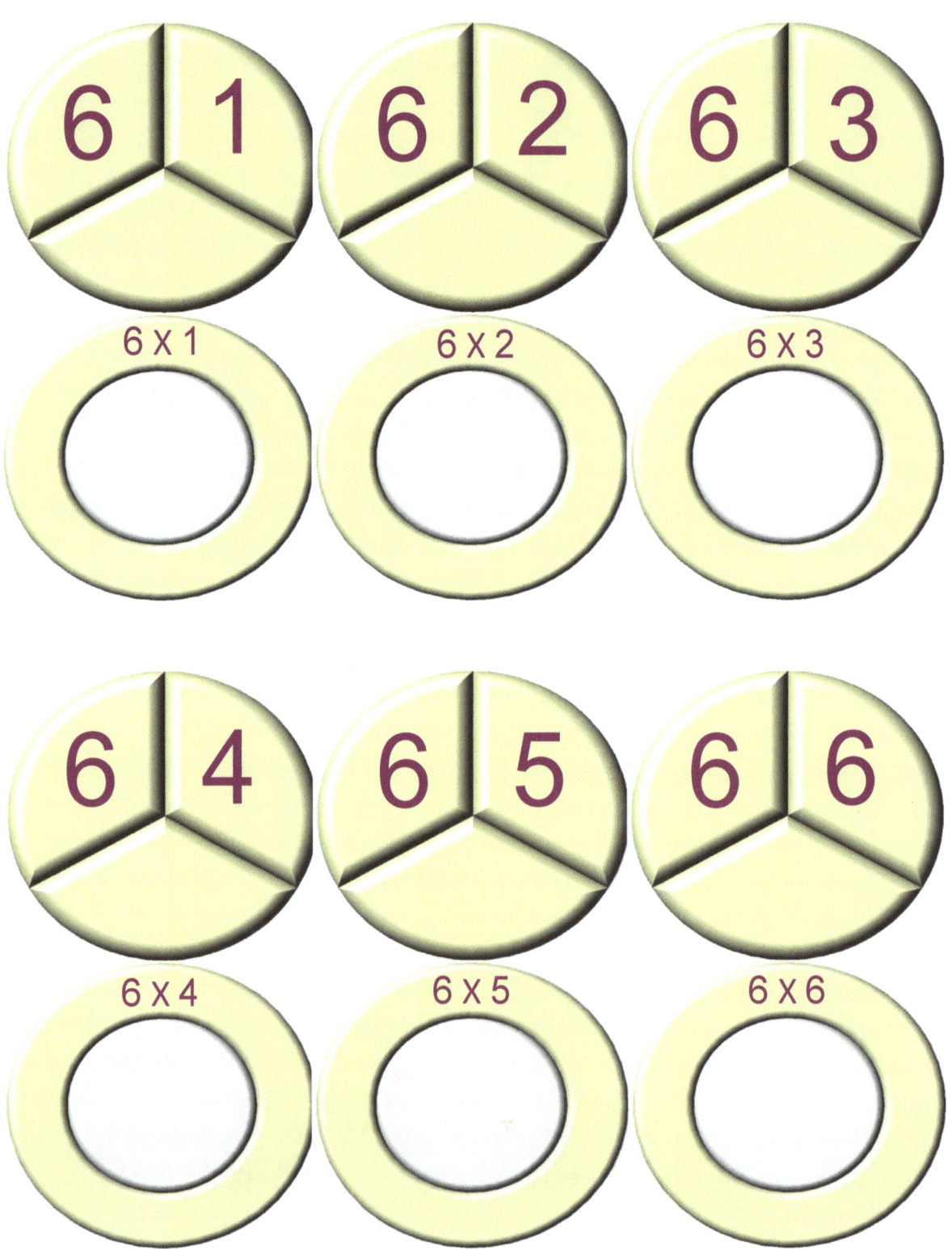

# Six times drills

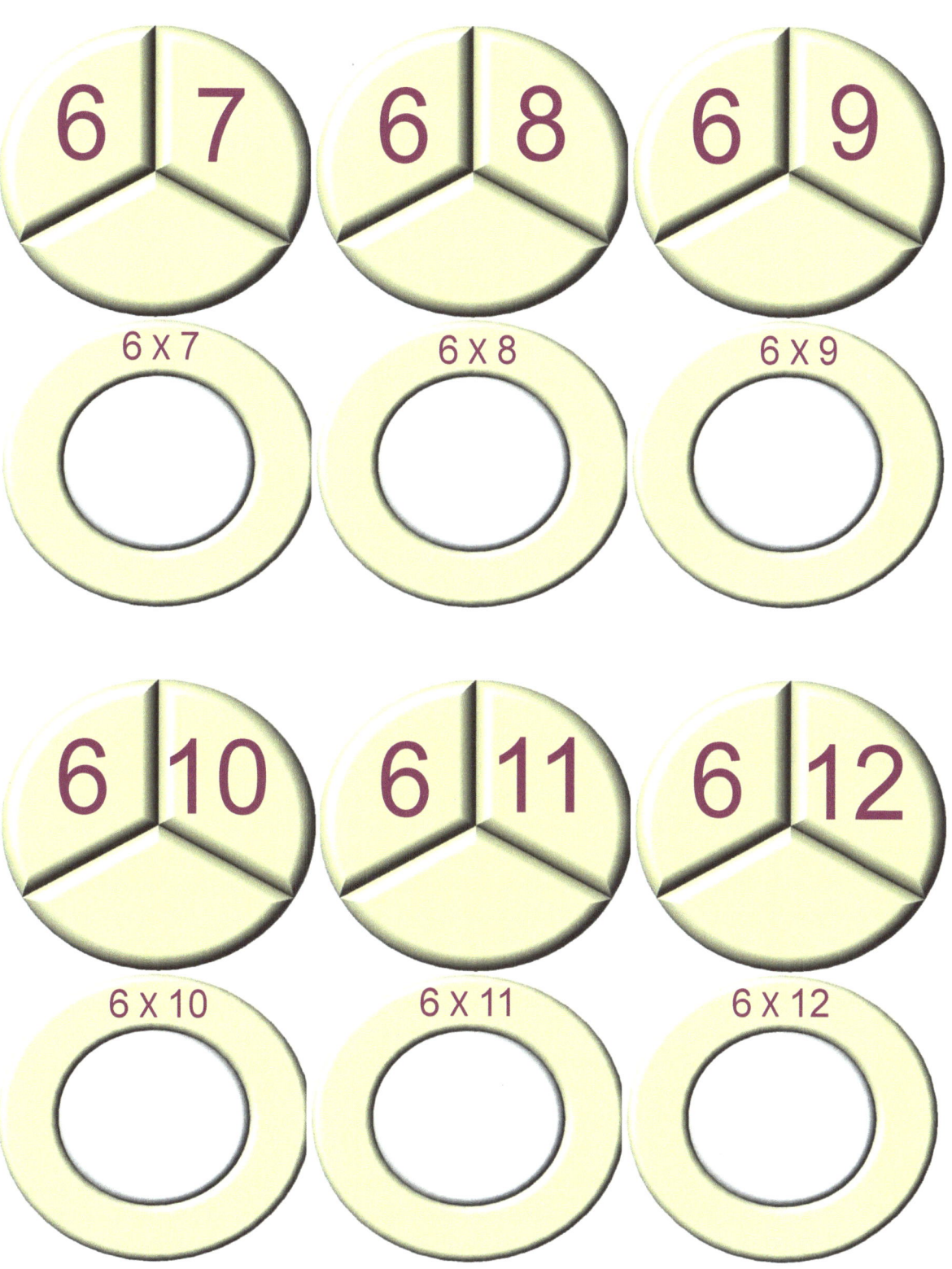

# Seven times drills

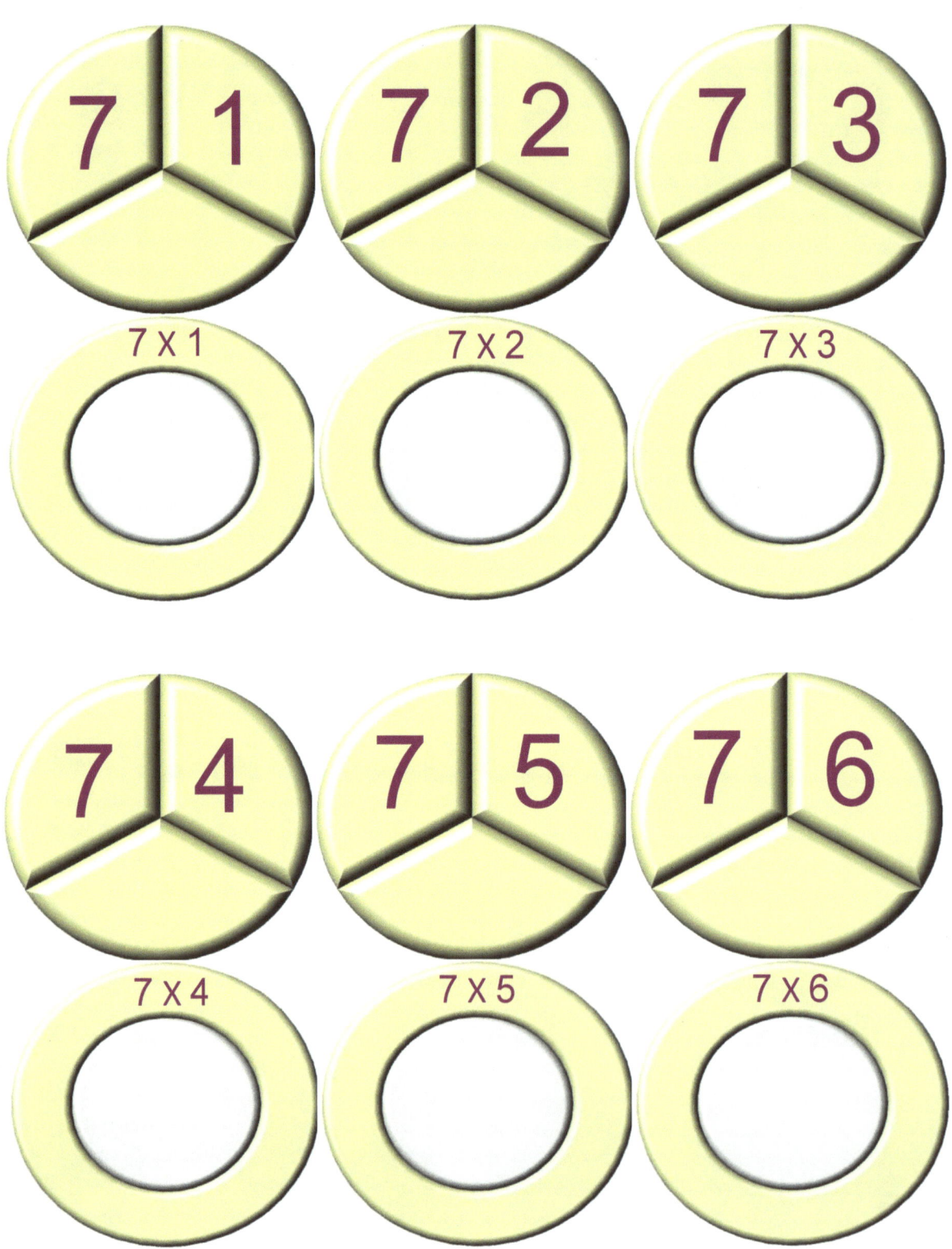

# Seven times drills

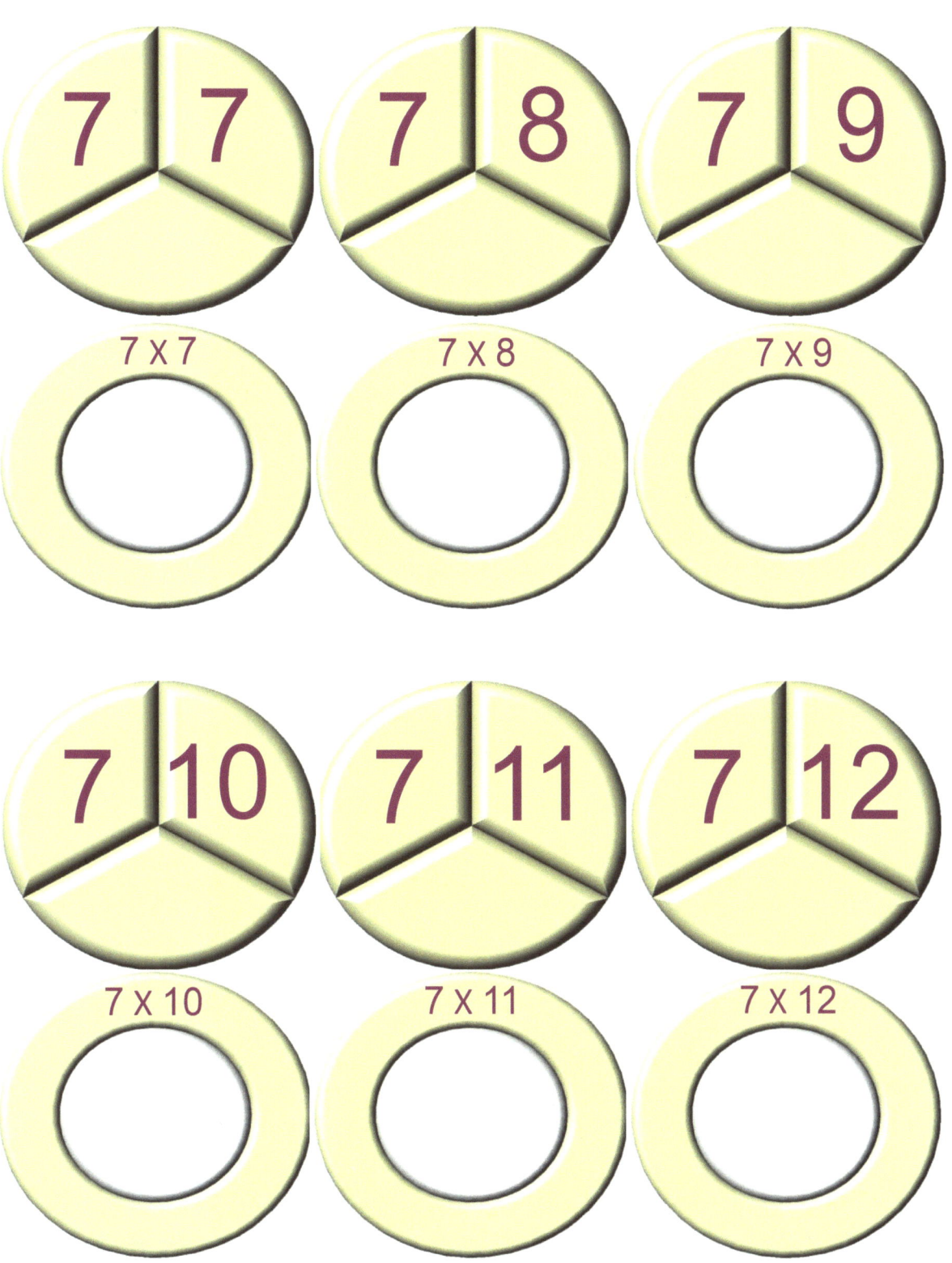

# Eight times drills

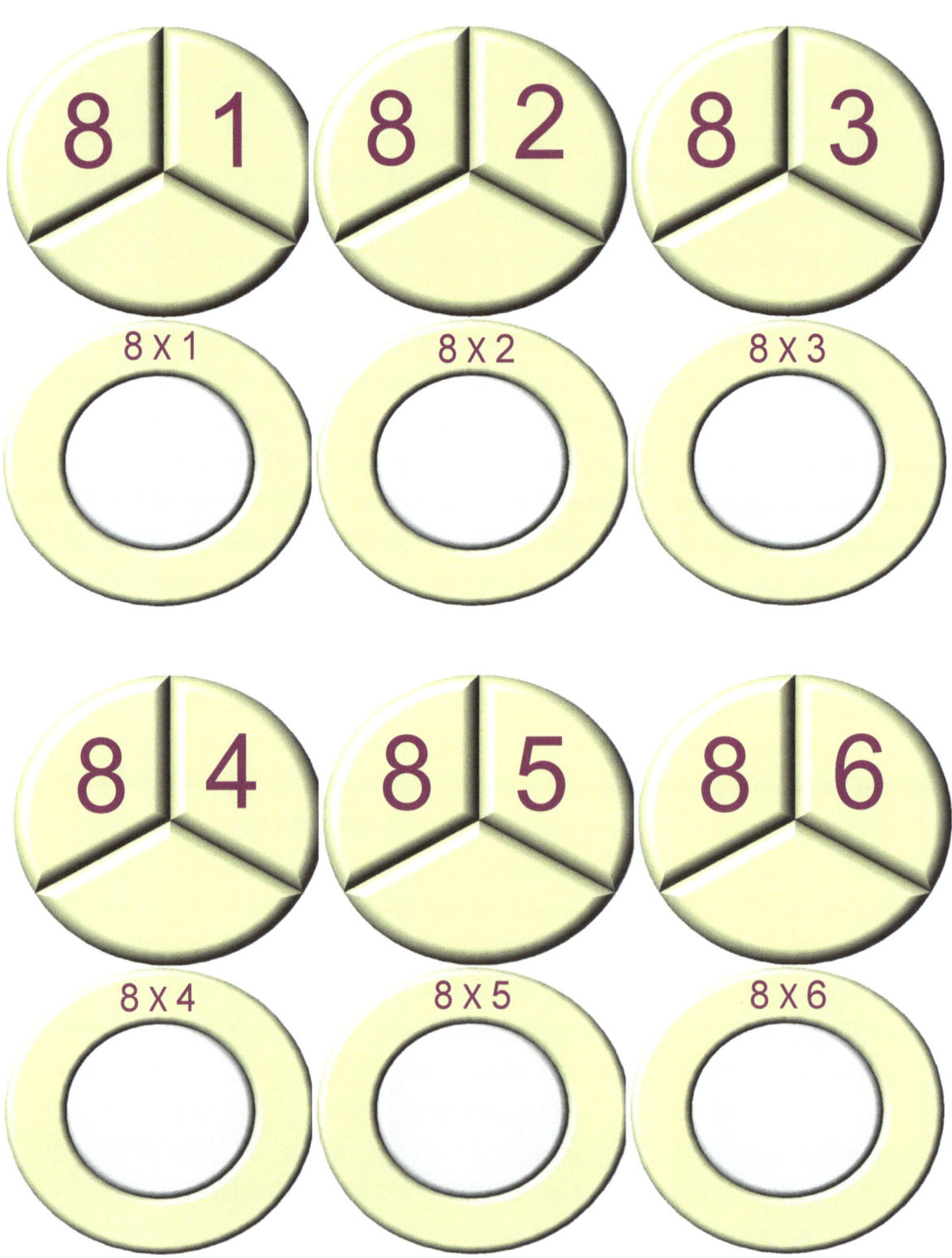

# Eight times drills

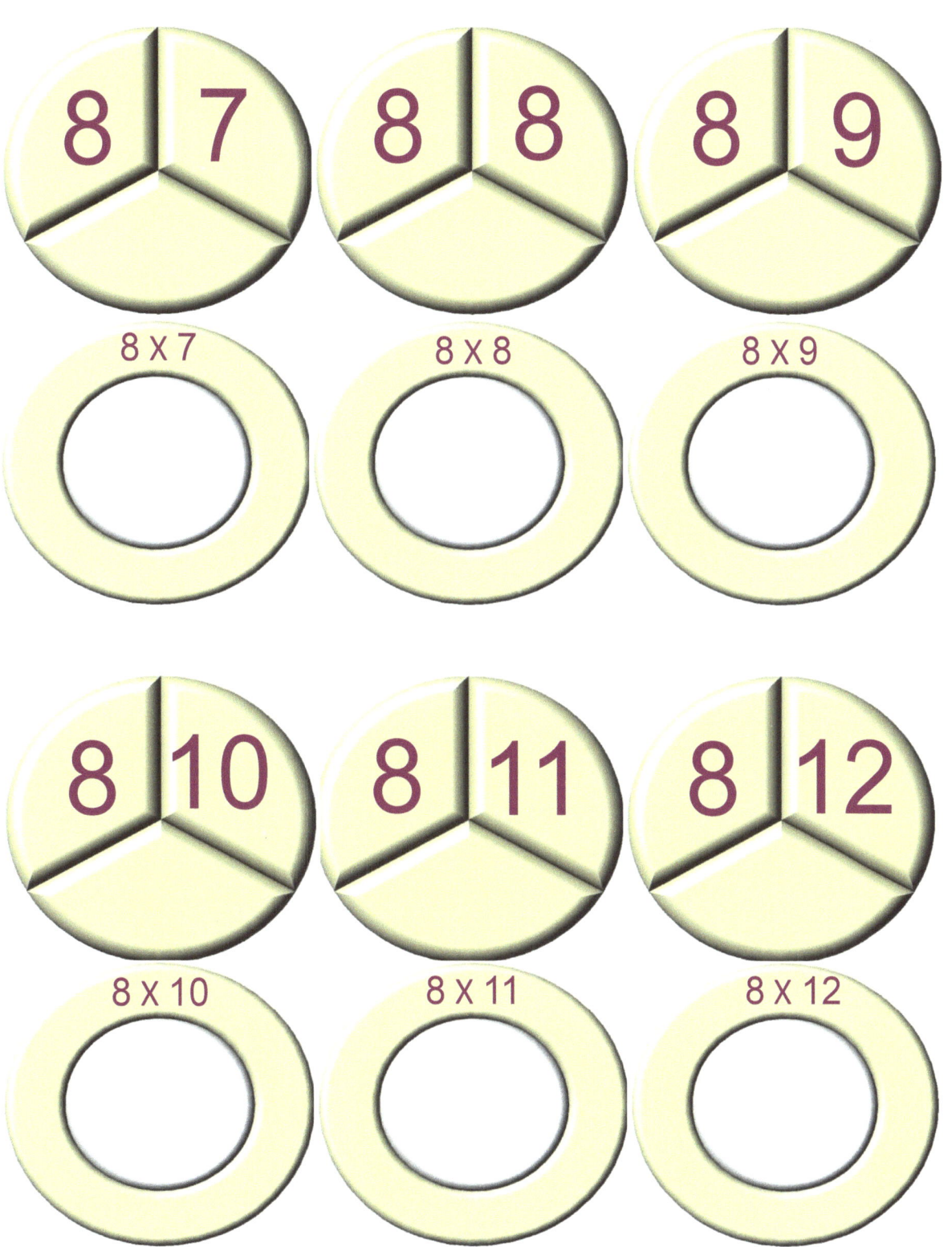

# Nine times drills

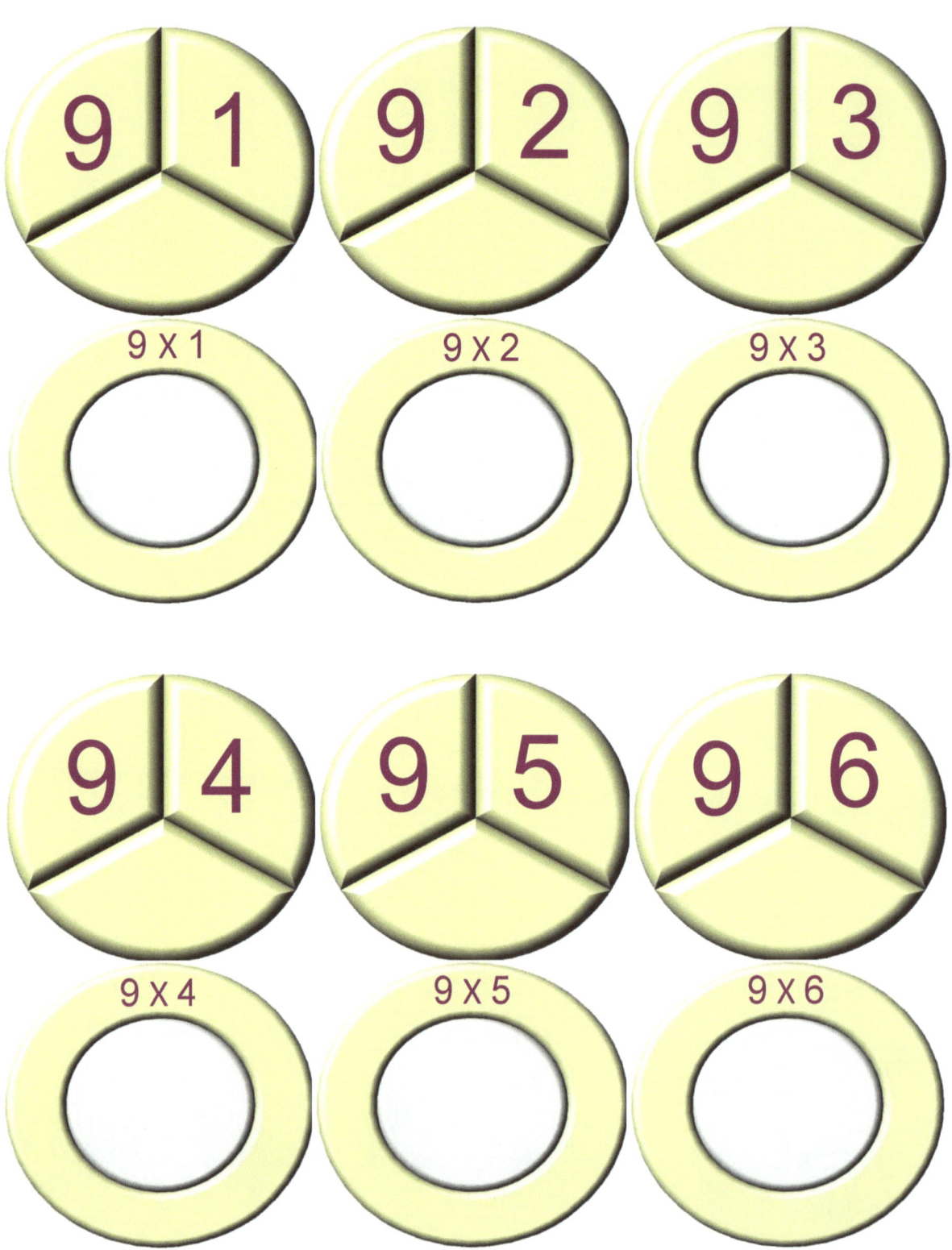

# Nine times drills

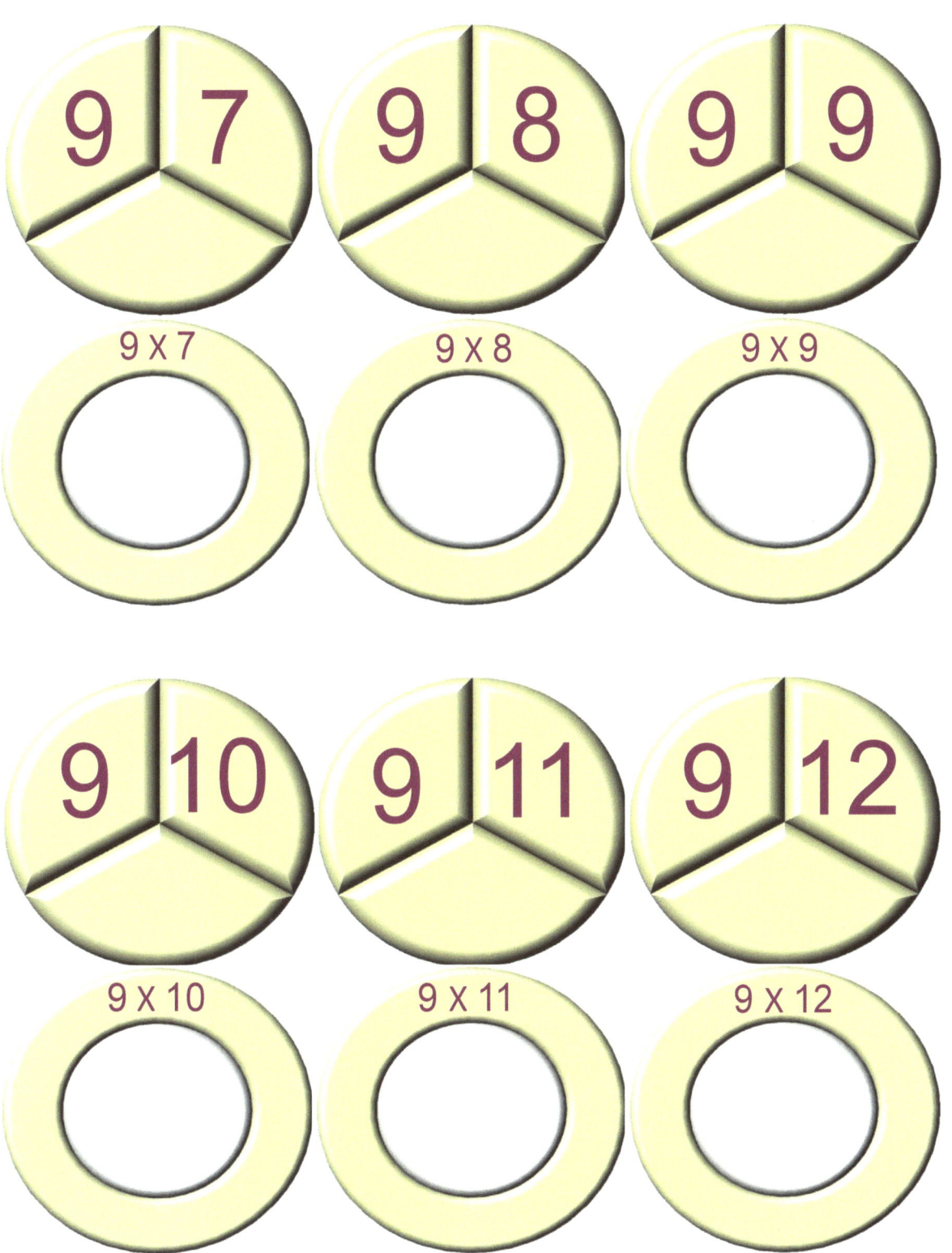

# Ten times drills

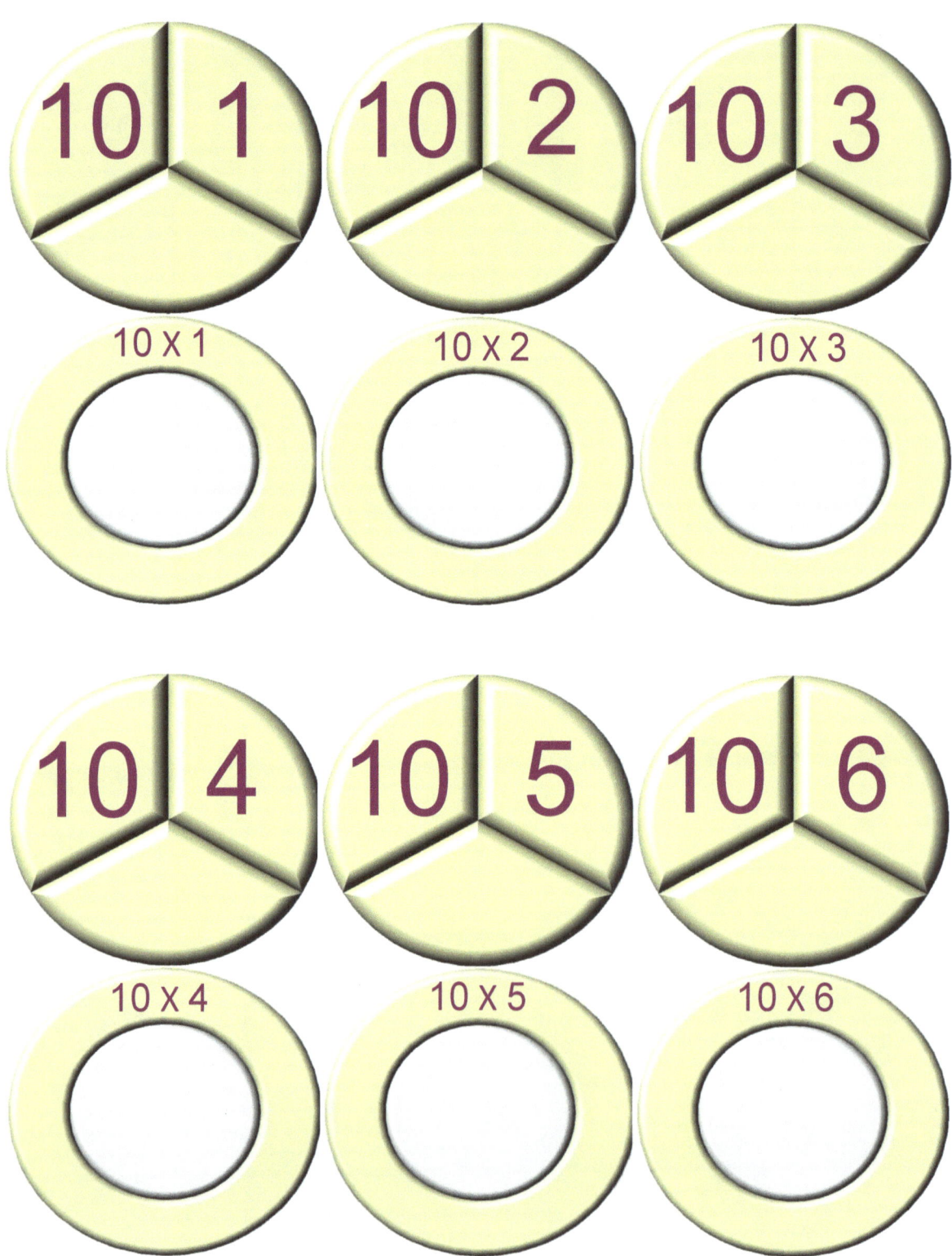

# Ten times drills

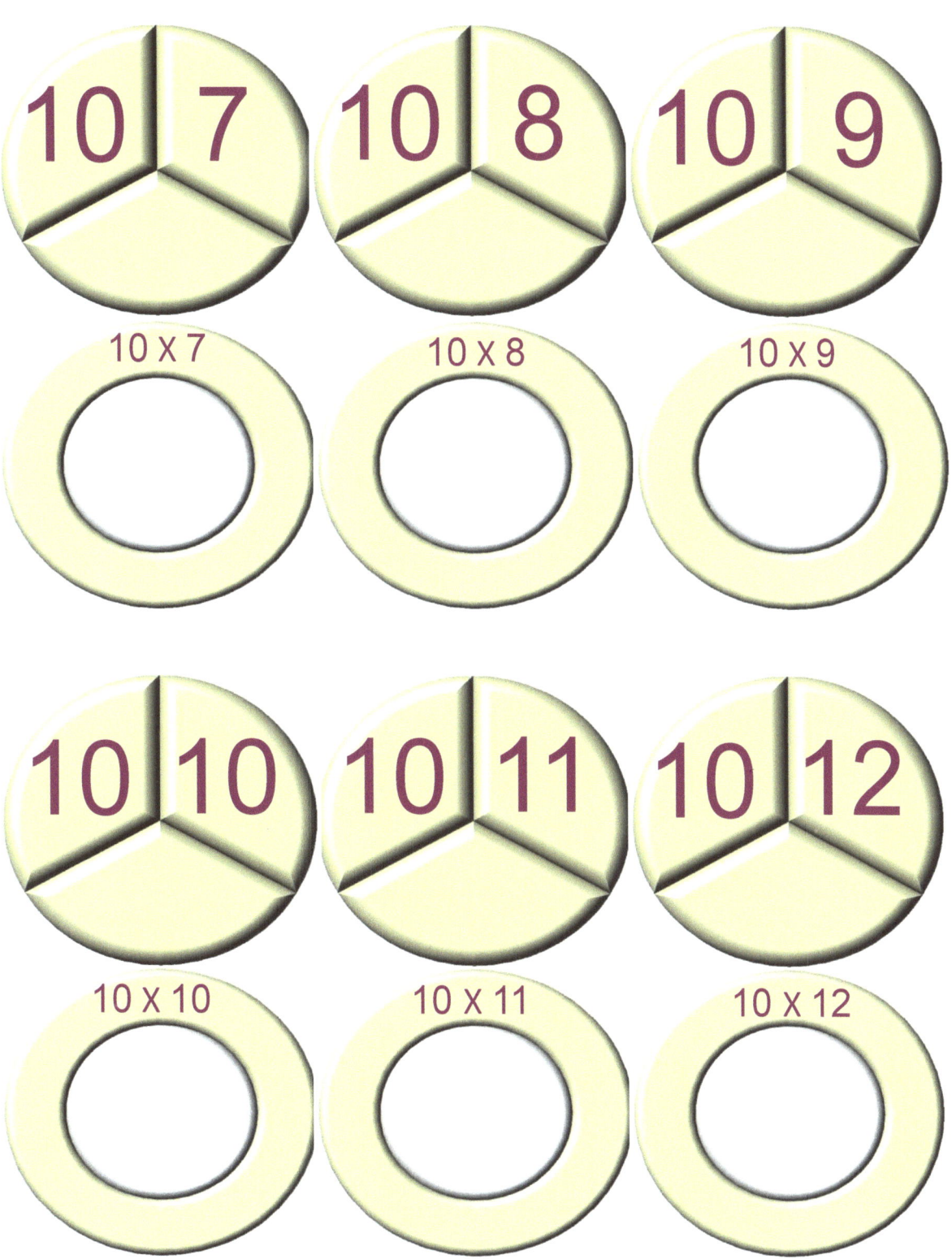

# Eleven times drills

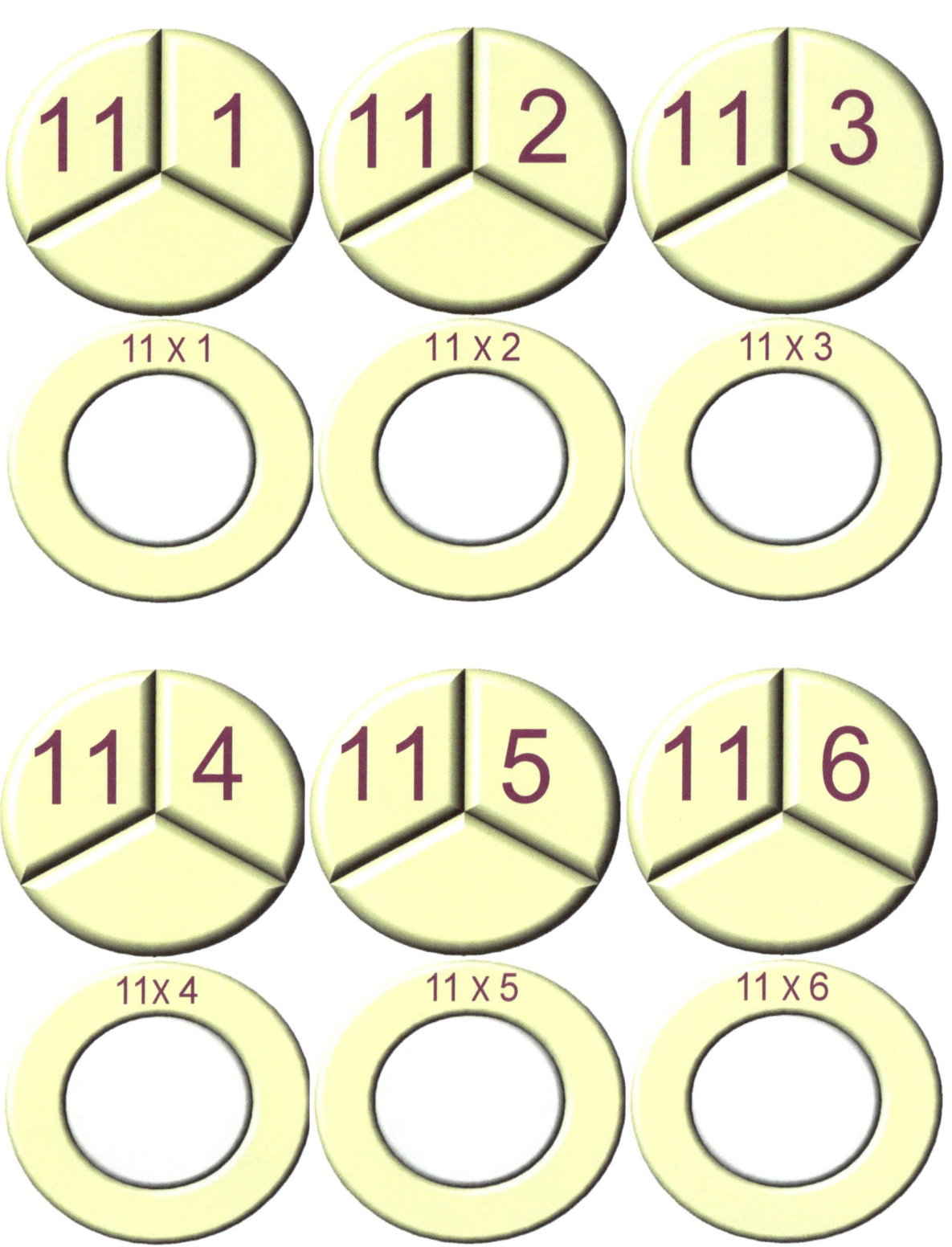

# Eleven times drills

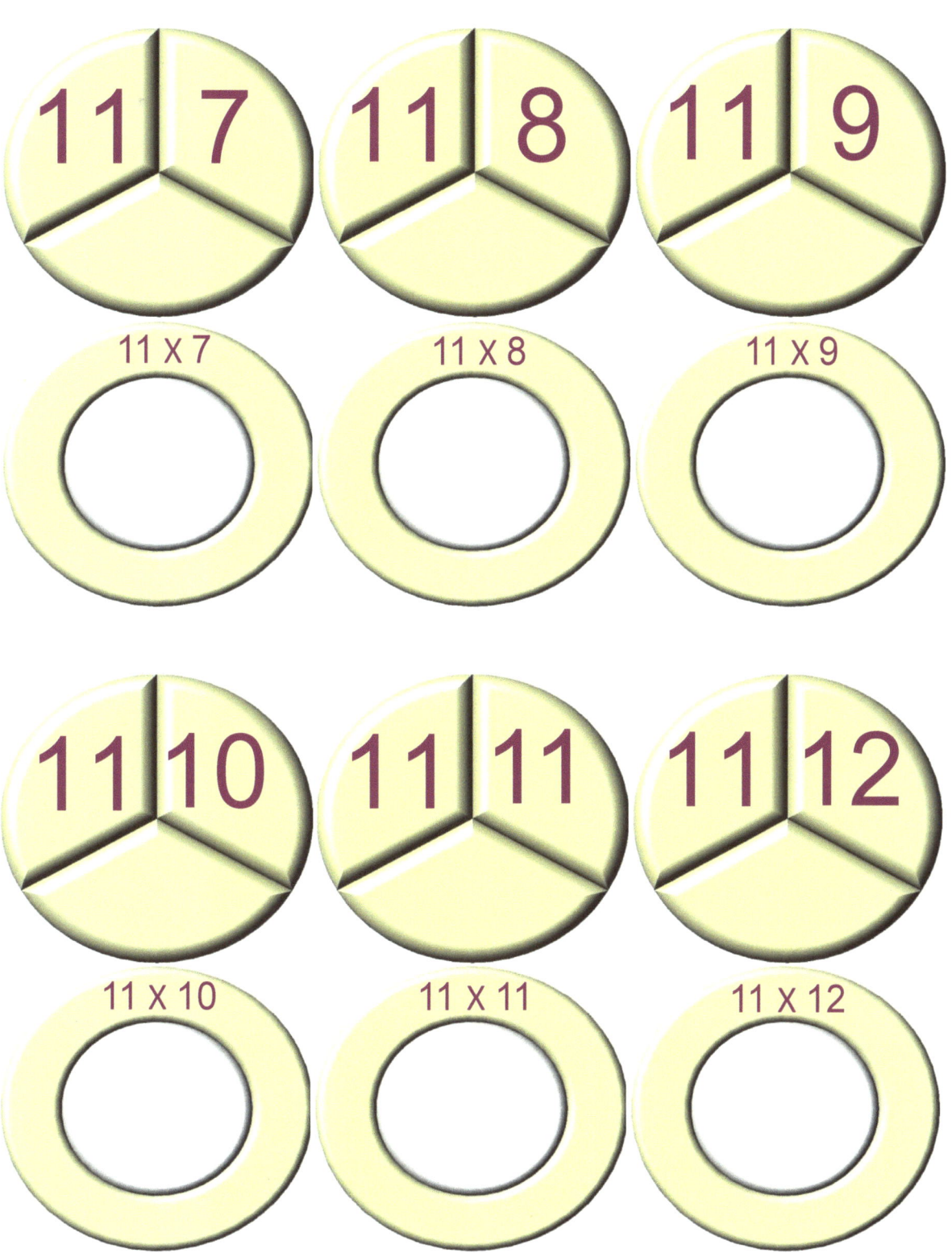

# Twelve times drills

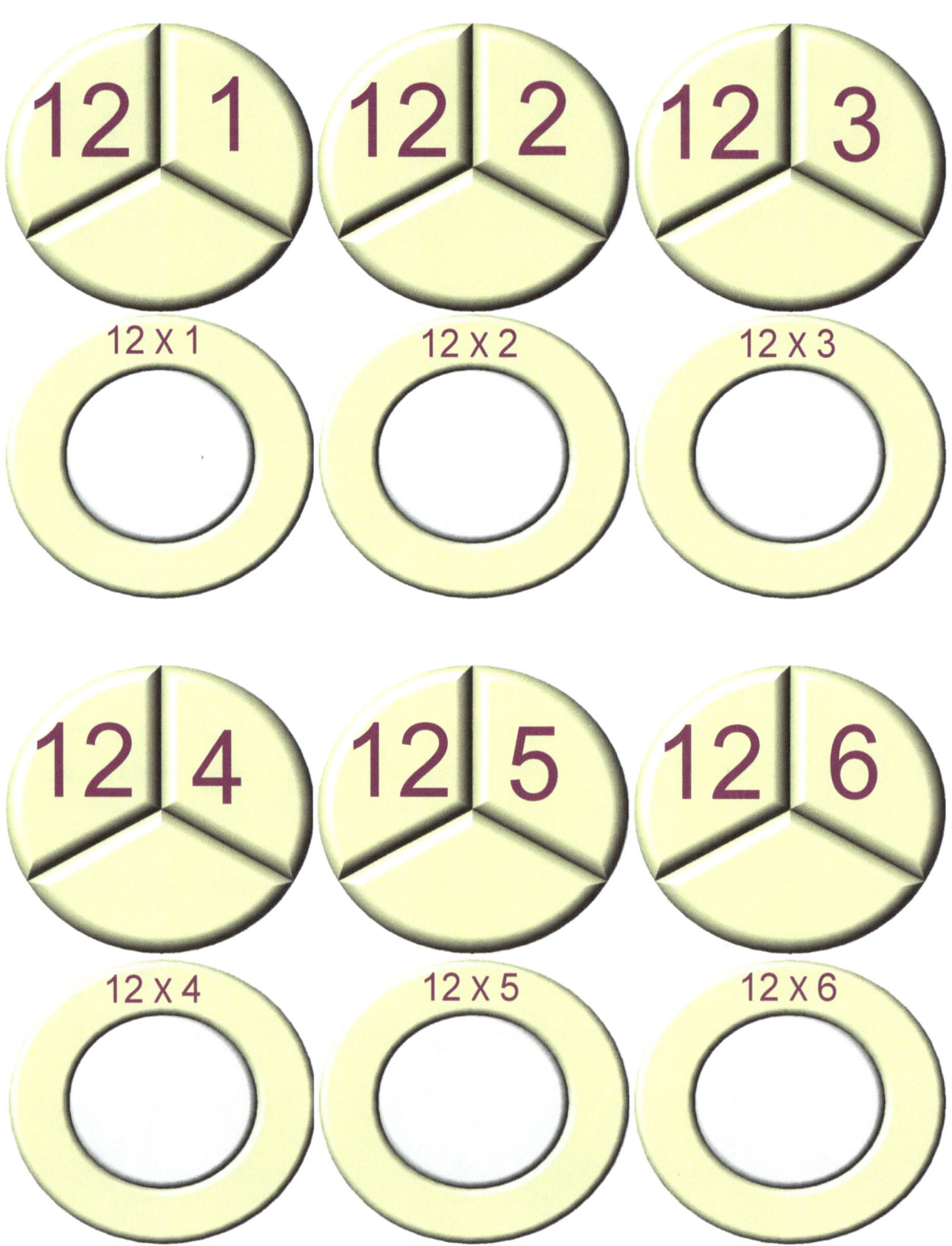

# Twelve times drills

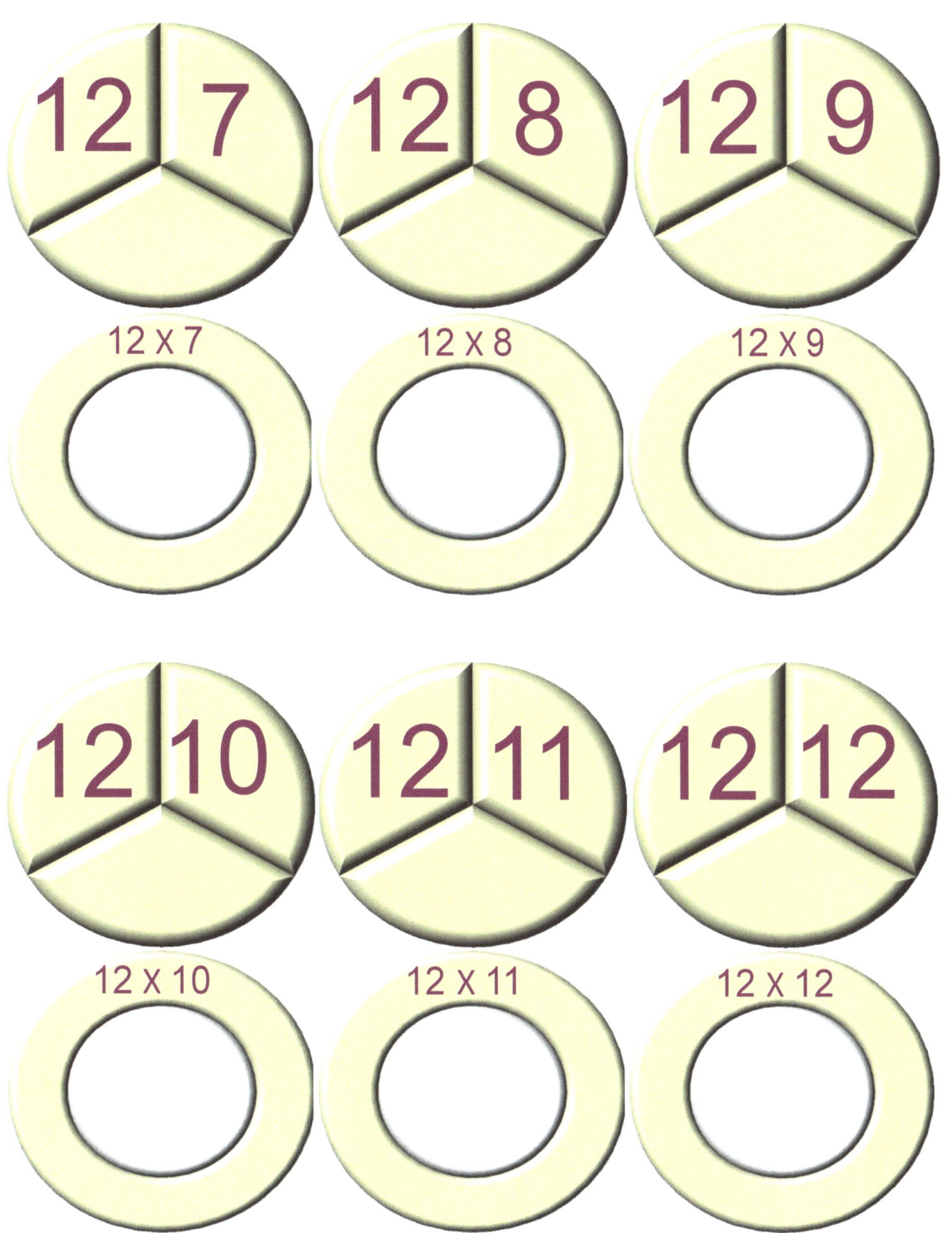

# One times drills

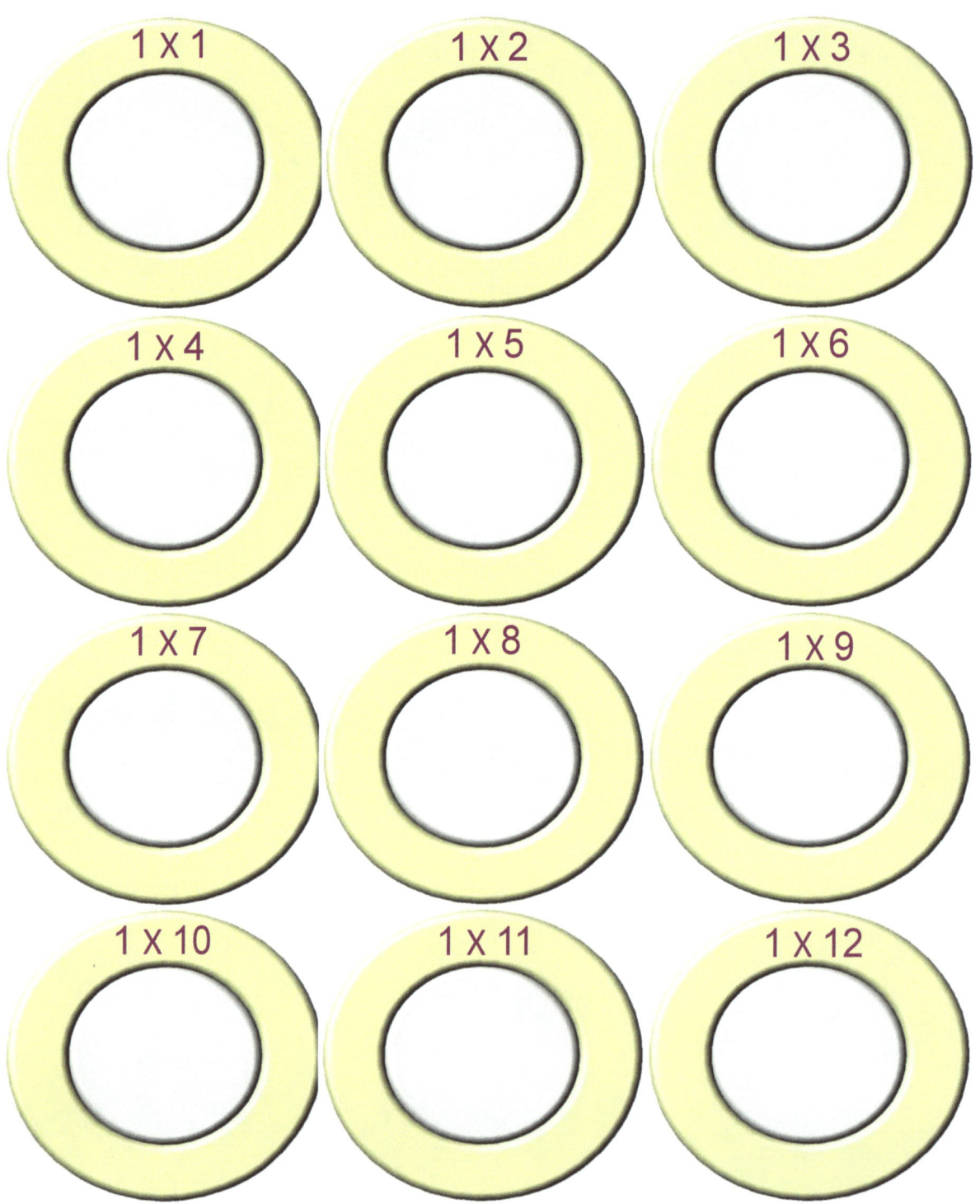

# Two times drills

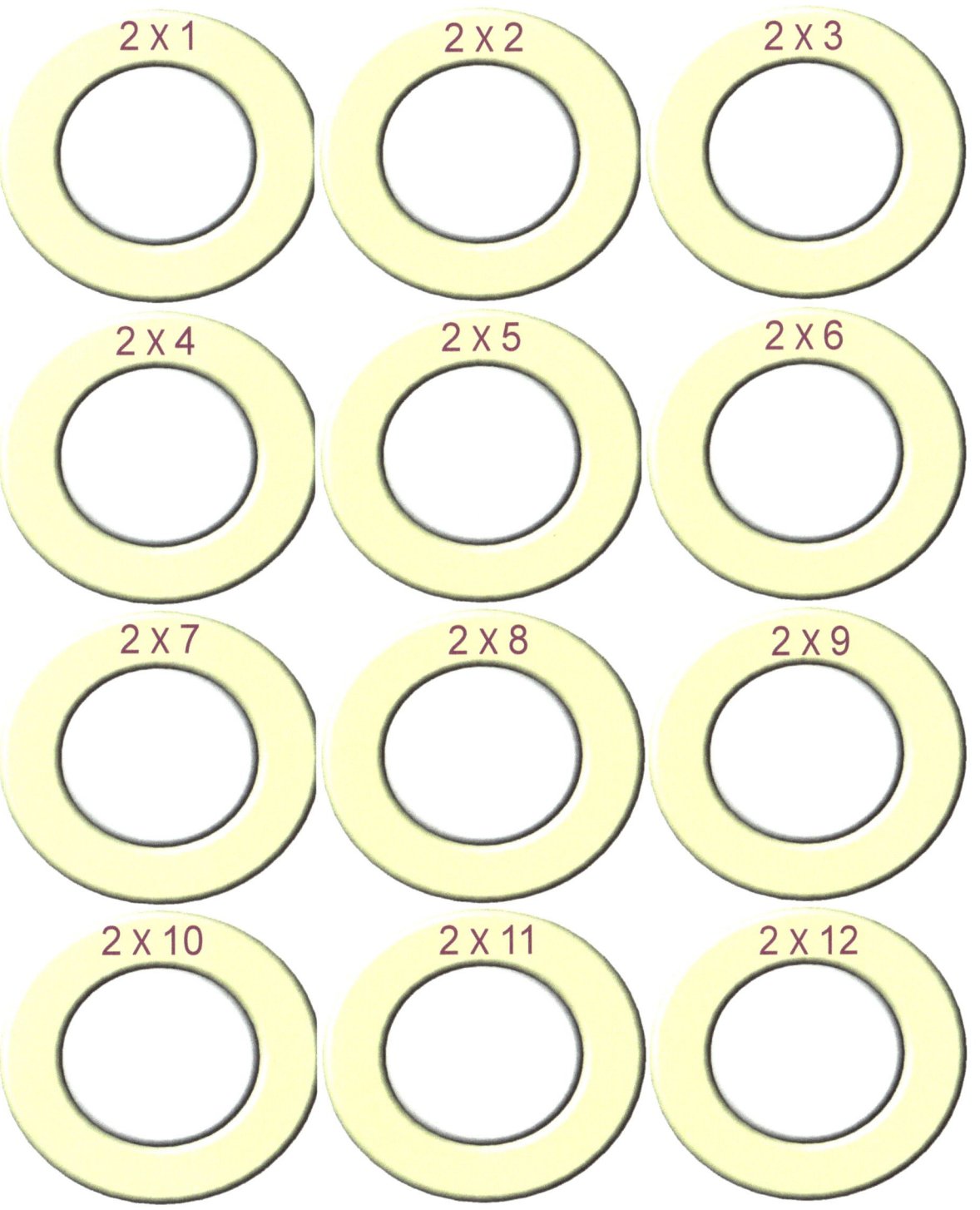

# Three times drills

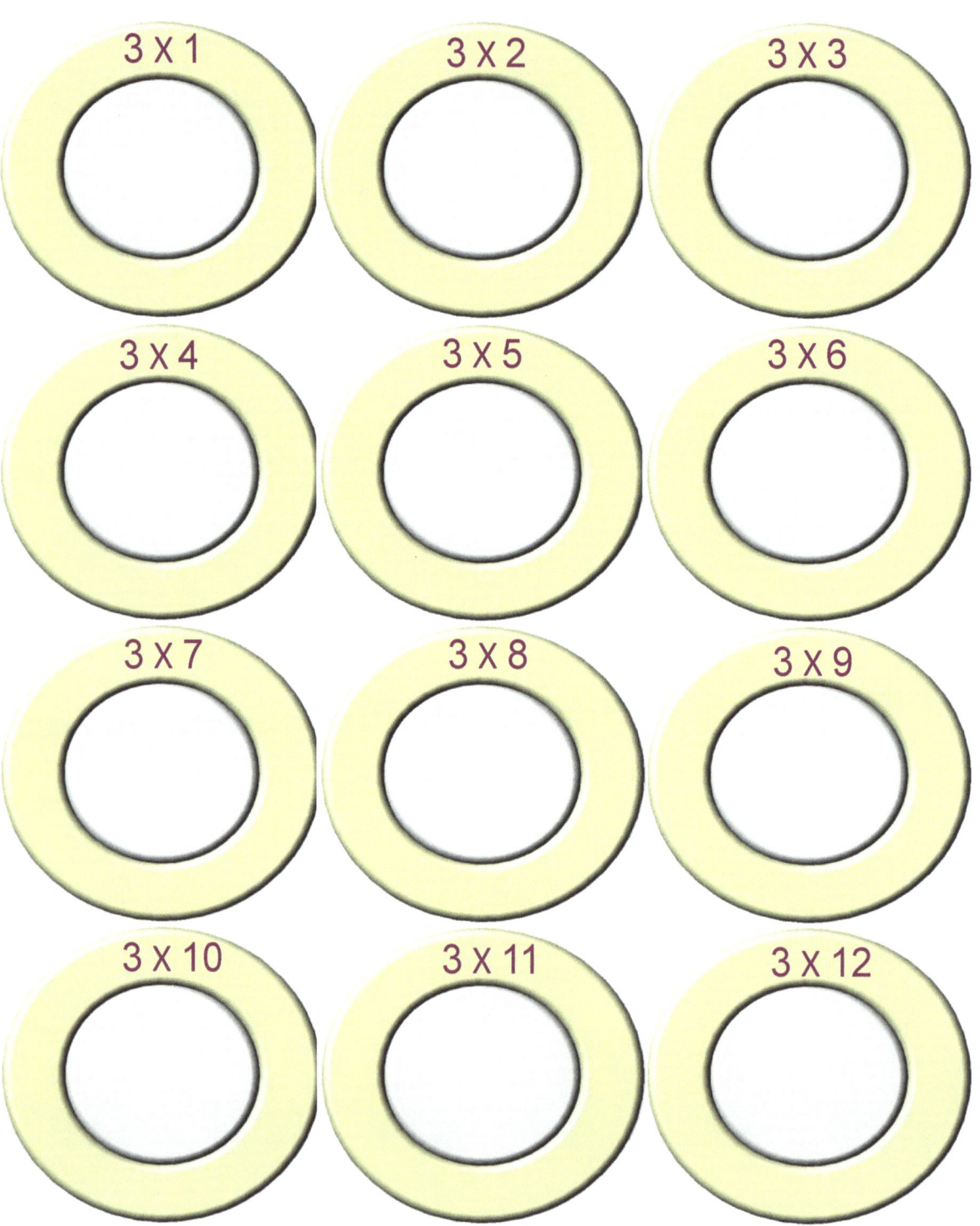

# Four times drills

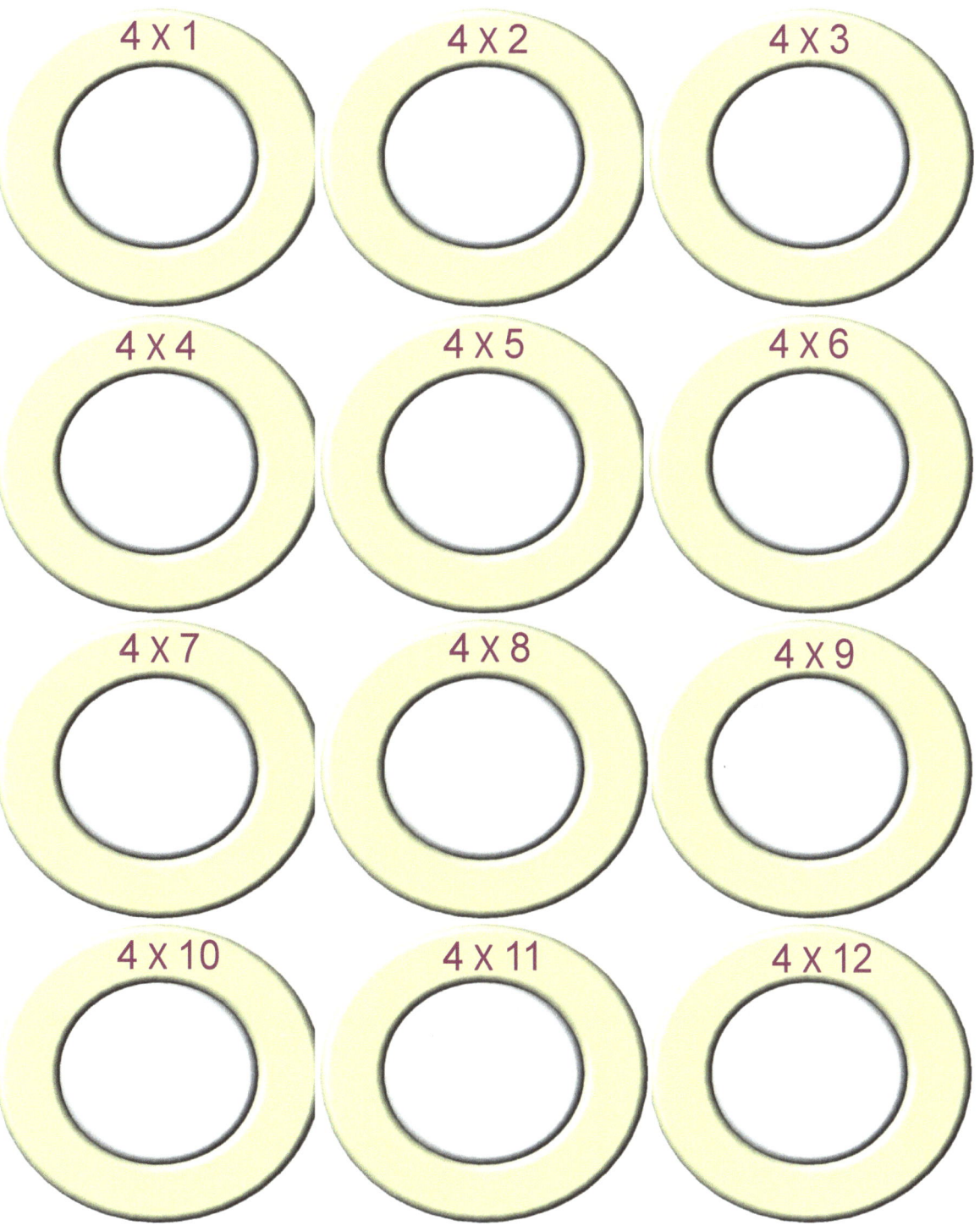

# Five times drills

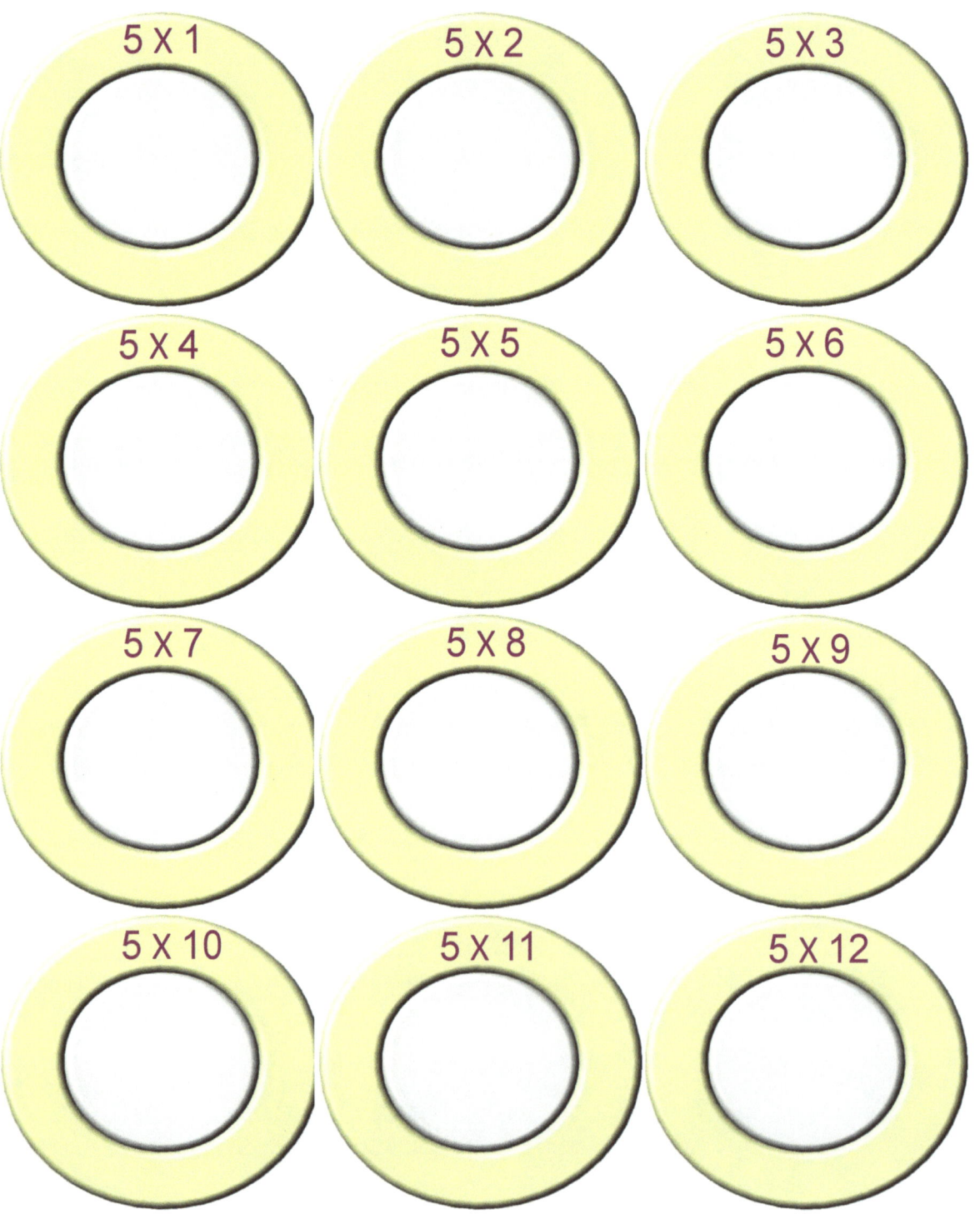

# Six times drills

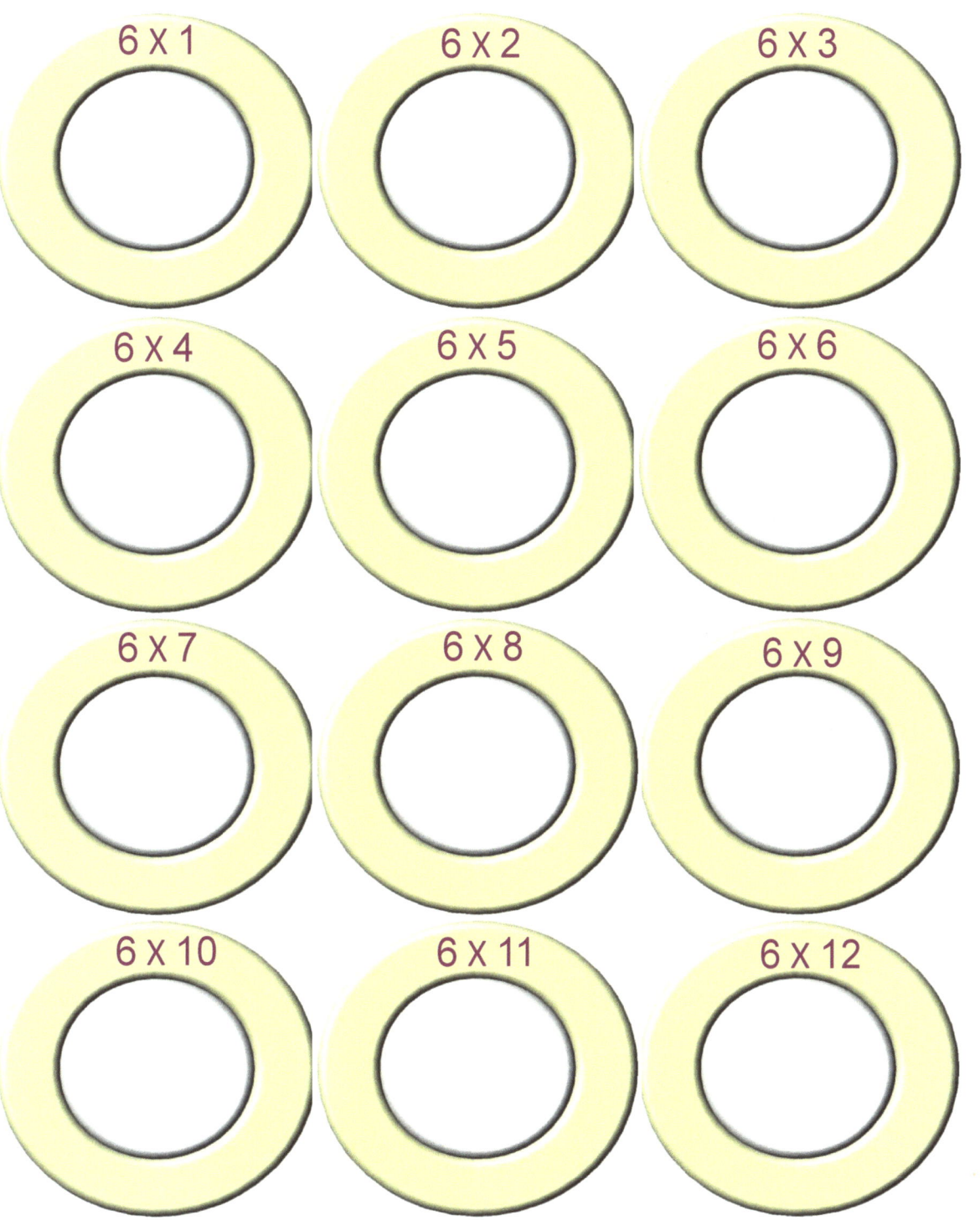

# Seven times drills

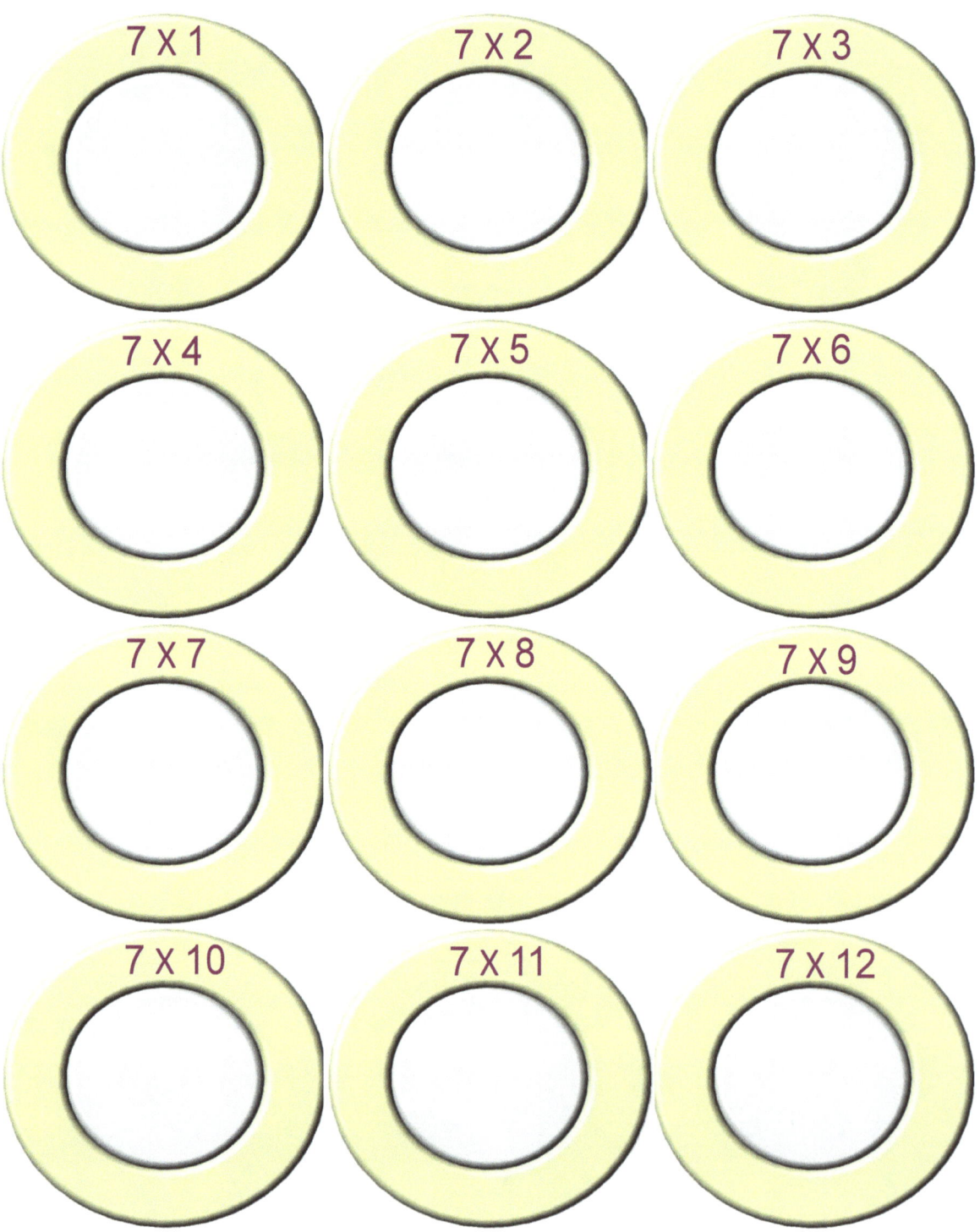

# Eight times drills

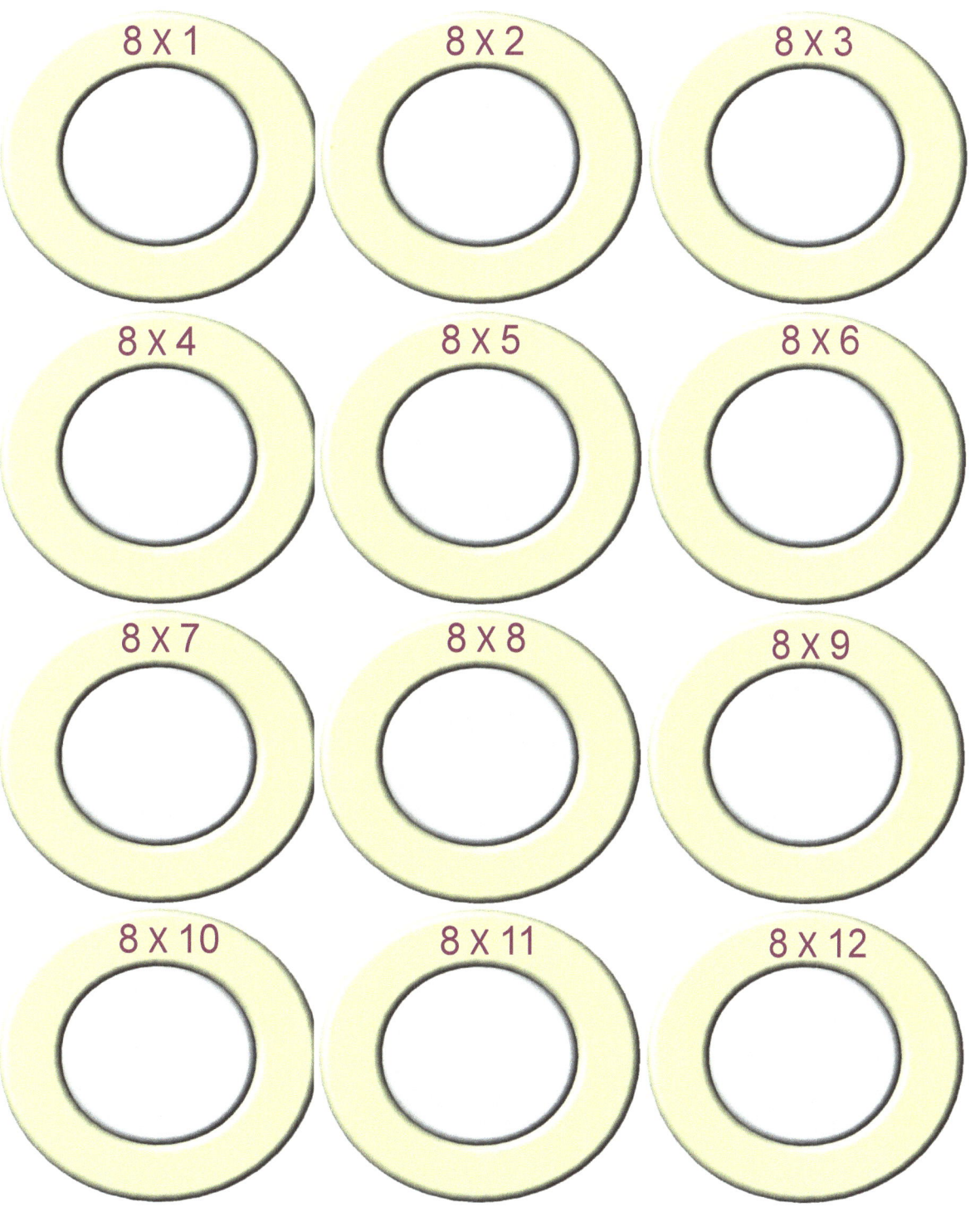

# Nine times drills

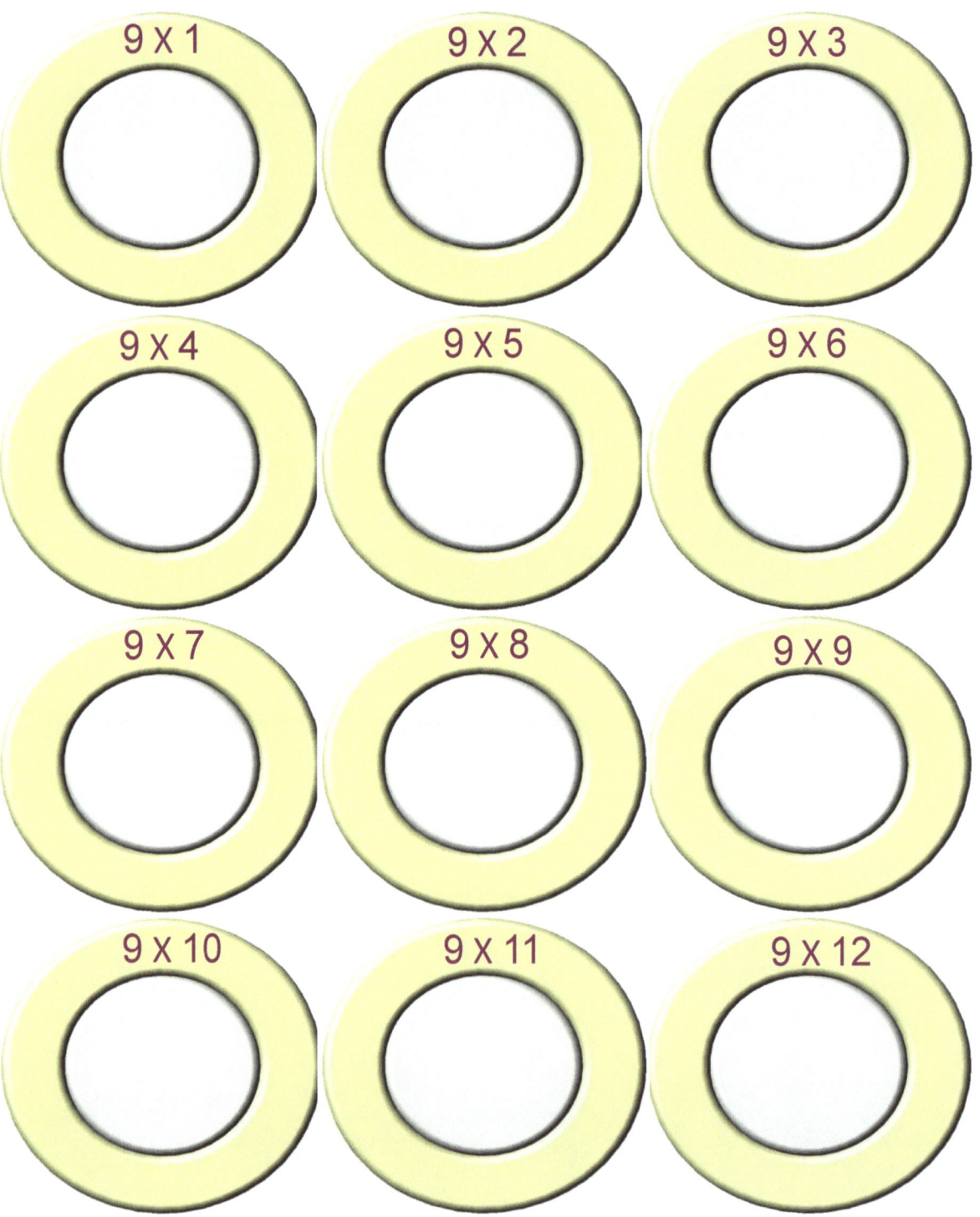

# Ten times drills

# Eleven times drills

# Twelve times drills

www.ingramcontent.com/pod-product-compliance
Lightning Source LLC
Chambersburg PA
CBHW051156220526
45473CB00003B/799